New Approaches to Characterization and Recognition of Faces

New Approaches to Characterization and Recognition of Faces

Editor

Asa Bensten

scitus
academics

New Approaches to Characterization and Recognition of Faces
Edited by **Asa Bensten**

ISBN: 978-1-68117-463-1
Library of Congress Control Number: 2016942825

www.scitusacademics.com
Box No. 4766, 616 Corporate Way,
Suite 2, Valley Cottage,
NY 10989

Printed in United States of America on Acid Free Paper

Preface

Man's face plays an important role in our social interaction and transmits the identity of people. Biometric face recognition technology has received considerable attention in recent years, due to its potential for a wide range of applications in both law enforcement and non-law enforcement, using the human face as a key to security. Compared to other biometrics systems using fingerprint /palmprint and iris, face recognition has distinct advantages due to its inaccessible process. Facial recognition (or face recognition) is a type of biometric software application that can identify a specific individual through the analysis and comparison of patterns in a digital image. Facial recognition systems are commonly used for safety purposes but are used in a variety of other applications increasingly. Most current facial recognition systems work with faceprint numeric codes. It identifies 80 nodal points on a human face. New ways of characterizing and recognizing faces are arranged around a number of clustered themes covering various aspects of face recognition.

This book explores several methods for the reduction of dimensionality in face image processing, statistical methods for face detection and smart face detection methods dominated by the use of artificial neural networking. This book will be of valuable tool for postgraduate students, practitioners, and researchers as well as those working in this field.

Table of Contents

Chapter 1

Fusion of Visible and Thermal Descriptors Using Genetic Algorithms for Face Recognition Systems

Gabriel Hermosilla [1,*], Francisco Gallardo [1], Gonzalo Farias [1] and Cesar San Martin [2]

[1]Escuela de Ingeniería Eléctrica, Pontificia Universidad Católica de Valparaíso, Valparaíso 2362804, Chile

[2]Department of Electrical Engineering, University of La Frontera, Temuco 4811230, Chile

ABSTRACT

The aim of this article is to present a new face recognition system based on the fusion of visible and thermal features obtained from the most current local matching descriptors by maximizing face recognition rates through the use of genetic algorithms. The article considers a comparison of the performance of the proposed fusion methodology against five current face recognition methods and classic fusion techniques used commonly in the literature. These were selected by considering their performance in face recognition. The five local matching methods and the proposed fusion methodology are evaluated using the standard visible/thermal database, the Equinox database, along with a new database, the PUCV-VTF, designed for visible-thermal studies in face recognition and described for the first time in this work. The latter is created considering visible and thermal image sensors with different real-world conditions, such as variations in illumination, facial expression, pose, occlusion, *etc.* The main conclusions of this article are that two variants of the proposed fusion methodology surpass current face recognition methods and the classic fusion techniques reported in the literature, attaining recognition rates of over 97% and 99% for the Equinox and PUCV-VTF databases, respectively. The fusion methodology is very robust to illumination and expression changes, as it combines thermal and visible information efficiently by using genetic algorithms, thus allowing it to choose optimal face areas where one spectrum is more representative than the other.

KEYWORDS

face recognition; fusion descriptors; genetic algorithms; visible and infrared spectrum

1. INTRODUCTION

Face recognition in the visible spectrum is a well-known field of research with many years of study using low-cost cameras. Many publications are available, containing smart and sophisticated algorithms. This field has seen steady growth due to its security applications, as well as other types of applications, such as access permission and even identity control. However, one of the main problems that researchers must overcome is the diminished recognition ability of algorithms as a result of variations in the intensity of illumination in the image. In a real application, where intensity of illumination can vary due to factors, such as the time or weather, face recognition with visible imagery may no longer be robust (*i.e.*, low recognition rates).

A solution to the problem caused by lighting in visible imagery is to use images acquired in the infrared (IR) spectrum (specifically, the thermal spectrum, 8–12 µm), because the images obtained in this spectral range remain invariant to lighting changes that may occur in the visible spectrum [1]. This is possible because thermal cameras capture the infrared radiation emitted by the face instead of the reflected light, following Stefan Boltzmann's law. The estimation of the temperature of a body is a process called thermography [2], in which it is not necessary to make contact with the surface of the body.

A disadvantage of the IR spectrum is that the energy received by the IR camera not only depends of the measured body, but also on the sum of the energy components of the different elements of the scene captured by the camera. The elements of the scene that are captured are: the atmosphere, the background and the object to be measured. The variation of these components could affect the temperature estimation of the measured body, faces in this case, which is a problem in face recognition. The thermal face images also have undesirable variations due to changes in ambient temperature, alterations of the metabolic processes of the subjects, camera susceptibility to extrinsic factors and variable sensor responses over time when the camera is working for long periods of time [3,4].

Based on the advantages and disadvantages of both spectra, it is possible to create a face recognition system based on the fusion of the visible and thermal domain, achieving a high recognition rate. The aim of this article is described in the following three points: (1) the article proposes a new fusion system that combines visible and thermal features obtained from the most current local matching descriptors by maximizing face recognition rates through the use of genetic algorithms (GA, [5,6]); (2) the article compares and studies current face recognition methods and classic fusion approaches against the proposed fusion technique; and (3) the PUCV-VisibleThermal-Face (PUCV-VTF) database is described for the first time for face recognition, as it was designed for visible-thermal studies in face recognition. The main idea of the fusion method is to maximize face recognition rates using genetic algorithms, which improves recognition rates using visible and thermal descriptors obtained from the newest and best methods currently used in the literature.

The PUCV-VTF database contains thermal and visible images acquired with two different sensors and allows us to study the fusion of these images. The visible images were acquired with a PS3Eye camera, 640 × 480 pixels, while the thermal images were acquired with an FLIR Tau 2 640 camera, with a resolution of 640 × 512 and 14 bits.

This study includes five local matching methods obtained from different comparative studies of visible and thermal methods [7,8,9,10,11,12], such as local binary pattern (LBP) [13], the Weber law descriptor (WLD) [14] and the Gabor jet descriptor (GJD) [11], due to their high performance. In addition, new methods that work in the visible spectrum, such as local derivative patterns (LDP) [15] and HOG descriptors [16], are also included due to the high performance seen in recent studies [15,16,17]. To perform the fusion method and based on results obtained previously [7,8,12], we choose two local descriptor methods, the LDP and LBP descriptors, to perform the combination. Thus, the genetic algorithm combines both descriptors in order to maximize the recognition rate of the entire face recognition system.

The main differences between our proposed fusion method and other fusion studies [18,19,20] are presented in the following paragraph. In [18,19], standard appearance-based methods are used together with genetic algorithms for the analysis and fusion of visible and thermal data. The studies use the classic eigenfaces approach for fitness evaluation, in which it is difficult to describe the face efficiently, unlike with our method, which uses new approaches for face description. In addition, our way of performing the fusion is different, because we use only 64 weights for the fused descriptor against the 100 principal components in [18], for example. In [20], the fusion of visual and thermal images using the discrete wavelet transform (DWT) domain is described. The results of the experiments demonstrate that the fusion method is effective in terms of visual quality compared to conventional visible and thermal approaches, but it depends on glasses being present in the image. In our work, the selected face recognition descriptors are robust to the occlusion produced by glasses [7]. In [21,22], genetic algorithms are not used to perform the fusion; however, those articles use methods based on data fusion and decision fusion. In [21], the fusion algorithm is designed to detect and replace glasses with an eye template in the case of thermal images, and FaceIt, a commercial face recognition software, is used to evaluate the fusion algorithm. In [22], the advantages of combining thermal and visible face recognition are analyzed, and recognition is achieved using a k-nearest neighbor classifier by principal component analysis (PCA). Finally, [23] uses a fusion method based on image fusion using the entropy information with a genetic algorithm. The study does not perform face recognition; however, it shows a way to efficiently combine visible and thermal features to be used in a face recognition system.

The proposed fusion methodology and the current visible-thermal methods will be evaluated using the Equinox database [24] and a new database named PUCV-VTF, which has been created to compare recognition rate results and to validate the fusion method. The Equinox database was used in [7,9,10], and it consists of face images in the visible, long-wave infrared (LWIR), mid-wave infrared (MWIR) and near-wave infrared (NWIR) spectra. For this study, only the visible and LWIR are used with six gallery sets and nine test sets for both spectra. In addition, we will compare the face recognition results obtained using the fusion approaches from [18,19,20] against the proposed methodology using the Equinox database. The PUCV-VTF database consists of face images in the visible and long-wave infrared spectra (thermal images), taken at the same time. We will create five sets, using a frontal set as the gallery and the others as test sets. Specifically, in our experiments, we use the Equinox database to generate the weights with the genetic algorithm for the fusion method, and we then evaluate the fusion method using the PUCV-VTF database.

This paper is structured as follows: Section 2 describes the current visible and thermal methods used in face recognition systems. The proposed methodology of visible-thermal descriptor fusion using a genetic algorithm is explained in Section 3. The visible and thermal databases and the image sensors used in this study are presented in detail in Section 4. Section 5 shows the experiments to obtain the optimal population of the fusion scheme and the validation of the fusion methodology. Finally, the main conclusions are presented in Section 6.

2. VISIBLE-THERMAL METHODS

As was mentioned in the previous section, we selected five current methods used in the literature to generate face recognition systems in the visible and thermal domains. The methods were chosen considering high performance and requirements, such as working in real time, using only one image per person in a gallery for face recognition matching (no training stage) and their high rates in comparative studies [7,8,9,10,11,12].

2.1. LDP Histograms

The local derivative pattern of n-th order was first proposed in [15]. The method obtains micropatterns that use the information contained in directional derivatives of (n − 1) order of the original image. The method uses different $\boldsymbol{\alpha}$ directions to obtain a descriptor: 0°, 45°, 90° and 135° only, since 180°, 225°, 270° and 315° are implicitly contained in the directions used. The micropatterns are obtained using a neighborhood of 8 pixels around a central pixel, taken from the derivative image in the $\boldsymbol{\alpha}$ direction to extract the information using a comparative function. To give a holistic feature to LDP, the image is divided into square regions where histograms with 256 bins are computed. The information taken from all of the directions for every region is concatenated for the final LDP n-th order descriptor of the original image.

2.2. LBP Histograms

The local binary pattern was originally proposed in [13]. It was first designed for the description of textures and was then adapted to face recognition. The method compares the intensity differences between the central pixel and its neighborhood in a 3 × 3 region to generate a binary code. One advantage of this method is its computational efficiency; this is important in image processing in real time. With this method, three levels of locality are obtained: pixel level (information about patterns obtained pixel to pixel), regional level (where groups of pixels give information about small regions of the face) and the holistic level of locality. A global description of the face is obtained concatenating the regional LBP extracted features using histograms by region.

2.3. Histograms of Oriented Gradients

Histograms of oriented gradients, proposed in [16], were used to obtain a human descriptor for pedestrian applications. The HOG algorithm uses the information contained in both the magnitude and orientation of the gradients presented in the original image. The face image is divided into*mxm* regions, where the histograms are calculated with 9 bins using the information on magnitude and orientation. These histograms are concatenated to obtain the final HOG descriptor of the original image, taking half of the precedent region in both directions to overlap the face information.

2.4. Weber Local Descriptor

The Weber local descriptor was first proposed in [14] and is based on the fact that the human perception of a pattern does not depend solely on a stimulus change, but also on its intensity. Specifically, WLD consists of two-dimensional histograms: differential excitations and orientations. The differential excitation component is a function of the radius between two terms: the first term is the relative difference in intensity of the current pixel against its neighbors, and the second term is the intensity of the current pixel itself. The orientation component is the orientation of the gradient of the current pixel. Thus, given an image, both components are used (differential excitation and orientation) to build and concatenate the WLD histogram.

2.5. Gabor Jet Descriptor

Gabor filters are used in face recognition because of their similarity to cell behavior in visual perception and of their robustness to change in illumination. The filters extract characteristics by means of the selection of frequency, orientation and scale. The jets are calculated at different points on a grid, and then, the Borda count method is applied to obtain a ranking and, thus, to generate the recognition. The implementation for this study was based on [11], in which several local feature representations, classification methods and combinations of classifier alternatives were compared. That system uses Gabor jet descriptors as local features, which are uniformly distributed over the images, one wavelength apart. In each grid position of the test and gallery images and at each scale (multiscale analysis), the Gabor jets are compared using normalized inner products. These results are combined using the Borda count method.

To summarize, we used five local descriptor-based methods, LDP, LBP, HOG, WLD and GJD, to perform the face recognition. For LDP, LBP and HOG, we used different divisions of the image to give them a holistic feature in partitions of 20 × 4, 8 × 4 and 16 × 16 regions, in order to find the best partition regions in our analysis. For WLD, we consider the results obtained from [7], which uses only 20 × 4 partitions. The recognition for the local descriptor method is performed using a nearest neighborhood with one of the three following similarity measures: Euclidean distance (EU), histogram intersection (HI) and chi squared (X2). For the case of the Gabor jet descriptor, the image is not partitioned, and the Borda count is used to compute the matching for the face recognition. In addition, the following notation is used to name each method and variation in this study: A_B_C, A describes the name of the method (LBP, HOG, GJD, LDP n-th and WLD), B shows the similarity measure (EU, HI or X2); C denotes the number of regions used (80 for 20 × 4, 32 for 8 × 4 and 256 for 16 × 16). In the case of GJD, only the Borda count is used, and the image is used as a whole, so its notation is simply "GJD".

3. PROPOSED METHOD: VISIBLE-THERMAL FACE DESCRIPTOR FUSION BY GENETIC ALGORITHMS

Genetic algorithms have been used since 1975 by applying Darwin's theory [5,6], and their use is inherently associated with maximizing or minimizing the result of a fit function by means of natural selection and mutation genetics, thus comprising a robust method of finding solutions that is not mathematically guided.

We propose the use of a fusion technique based on genetic algorithms, which combines the most relevant information from visible and thermal images with image descriptors. Our genetic algorithm finds a genetic code, which consists of weights for the visible and thermal descriptors to maximize the recognition rate as an objective function. The fusion method is performed by taking an input pair of test images (visible and thermal) to be compared against a pair in the gallery set, and following this, two face recognition descriptors are applied: the visible descriptor (VD) and the thermal descriptor (TD) (any local descriptor can be chosen). To combine the descriptors, histogram intersection is used as the similarity measure to perform the fusion. Figure 1 shows the fusion method.

Combinations of image descriptors were selected based on the results of [7,12,15]; for the visible spectrum, two descriptors were chosen: the third order LDP descriptor (LDP3), with 256 divisions in the image, and the commonly-used LBP; for the thermal case, only the LBP descriptor with 32 regions was used, as it is simple and obtains high results. The fusion method combinations are therefore based on: LDP-LBP and LBP-LBP for the visible and thermal descriptors. In addition, note that the LDP descriptor for visible image uses 256 regions, and the thermal image uses 32 regions; thus, to perform the fusion, we take groups of 8 sectors for the visible LDP descriptor. Therefore, the length of the genetic code given by the GA is of 64 weights, where the first 32 correspond to visible weights and the following 32 to thermal weights. Once the recognition rate was obtained, our fusion scheme maximized this recognition rate using the weights obtained from the GA process.

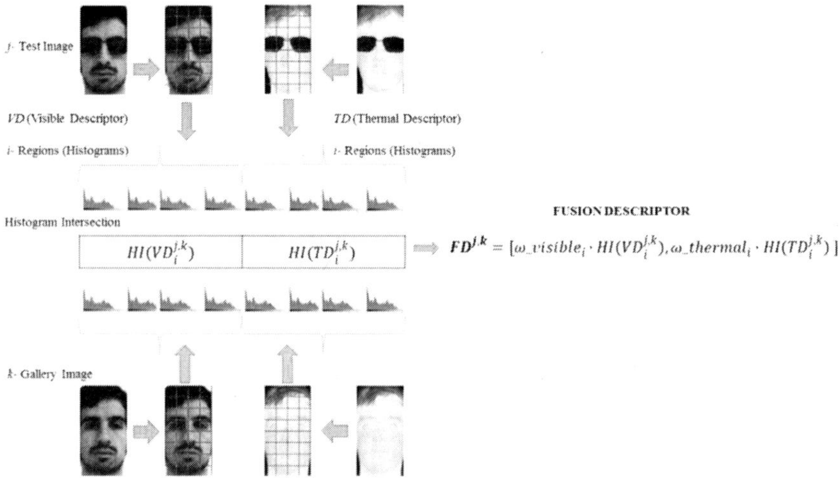

Figure 1. Fusion scheme process.

3.1. Proposed Genetic Algorithm for Fused Descriptors

Basically, a genetic algorithm (GA) searches for the optimum result of an objective function. First, an initial population of possible solutions has to be created and then evaluated using the objective function (also known as a fitness function) obtaining the fitness value of each individual in the population. After the initial population is established, several steps put the GA into a loop algorithm, which can be stopped by different criteria, such as an expected fitness value or number of iterations. This basic method also involves the selection of individuals as parents, crossover and mutation and finally reinsertion into the population.

The starting point of our GA was the choice of a function to maximize fitness, in our case the recognition rate. The GA assigned weights ω = [ω_visible ω_thermal] to the partitions of each image (visible and thermal), which represent the genetic code to be optimized. The process considers the face recognition results obtained from each test image *j*, gallery image *k* and the image partitions *i*of the regions of the histogram of the visible descriptor and the thermal descriptor (see Figure 1). Mathematically, the vector fusion descriptor (*FD*) for the test image *j* and the gallery image *k* can be described by the following Equation (1):

$$\mathrm{FD}^{j,k} = [\omega\_\mathrm{visible}_i \cdot \mathrm{HI}(\mathrm{VD}_i^{j,k}), \omega\_\mathrm{thermal}_i \cdot \mathrm{HI}(\mathrm{TD}_i^{j,k})] \quad (1)$$

where *i* is the number of regions of the image, *VD* is the visible image descriptor, *TD* is the thermal image descriptor and ω_*visible* and ω_*thermal* are the weights of the population to be optimized. The histogram intersection function *HI(·)* combines the descriptor regions from the test image *j* and gallery image *k* and is given by $\mathrm{HI}(S^{j,k}) = \sum_{w} \min(S_w^j, S_w^k)$.

Once the objective function was chosen and the nature of the genetic code was established, the GA was run with given gallery and test sets, aiming to find the best recognition performance between them using the GA with the following steps.

Initial population:

1 An initial population of 100 genetic codes was randomly made with 64 weights in the interval [0, 1]. The initial population had complementary weights for visible and thermal regions.

2 The genetic code was applied to the descriptor database of gallery and test images using Equation 1. The similarity value was then calculated using $\mathrm{SV}^{j,k} = \sum_{i=1}^{64} FD_i^{j,k}$.

3 The face recognition was performed, giving the maximum value of the similarity value (SV) to each test image j, and the recognition rate was then calculated using the correctly-recognized faces. Note that the fitness value of each genetic code is the recognition rate, which was then optimized. New offspring generation:

4 Two genetic codes were selected as parents, using the roulette method [5,6].

5 The parents were crossed with a probability of 25% at each point, obtaining two offspring (the weights of visible and thermal were no longer complementary).
6 The offspring was mutated at three random points with a probability of 25%.
7 The fitness values of both offspring were calculated as for the initial population (Steps 2–3).
8 If the fitness value of a given offspring was greater than the lowest fitness value in the population, then a random genetic code was replaced, otherwise it was discarded.
9 If the iteration number was under 100,000, Step 4 was repeated.
10 End.

The genetic algorithm process may take a considerable amount of time to obtain the genetic code. However, the GA process is performed only to find the best optimal weights to obtain the best genetic code, similar to an offline stage. After this process, the face recognition is performed with the fusion method using only the best code obtained.

4. VISIBLE AND THERMAL DATABASES

To carry out the fusion analysis, the Equinox database and the PUCV-VTF database were used. A brief description of the databases is given in the following sections. It is important to mention that the PUCV-VTF was created for this research and considers images in both the visible and thermal spectra of 76 individuals and different face conditions, such as emotions, gestures and artefacts (glasses).

4.1. Equinox Database Description

The Equinox database [24] has been used in several studies to compare and evaluate face recognition methods. Thus, it allows direct comparison between methods implemented previously, such as PCA, Independent Components Analysis (ICA), LBP, WLD, GJD, kernel PCA (KPCA) and kernel Fisher linear discriminant (KFLD) [9,10,19,25]. The Equinox database is often used to study the performance of different face recognition algorithms using thermal imaging (8–12 microns). The database consists of several sets of frontal images taken in both visible and infrared spectra at the same time, totaling 18,629 pictures for each spectrum with grey scale with a size of 240 × 320 pixels. The gallery set contains images from 81 different subjects, which were obtained under different lighting (frontal, left lateral and right lateral), three different expressions (smile, frown, surprise), while the test sets have more images per subject (see Table 1). Subjects who use glasses are captured twice, with and without their glasses. The images were processed to be 150 pixels in height and 81 pixels in width, with the eyes horizontally centered, aligned and with a distance between them of 42 pixels, showing only the face of the subjects (see the examples in Figure 2).

For the evaluation of the methods used in this study, 15 sets of images were established with the following criteria (taken from [12]). All subjects have images in each set (visible and thermal); for gallery sets, there is only one image per subject (81 images per set); the test sets have different sizes, as detailed in Table 1. Thus, six gallery sets were established (VA, EA, VF, EF, VL, EL) and nine test sets (VA, EA, VF, EF, VL, EL, VG, EG, RR), see details in Table 1.

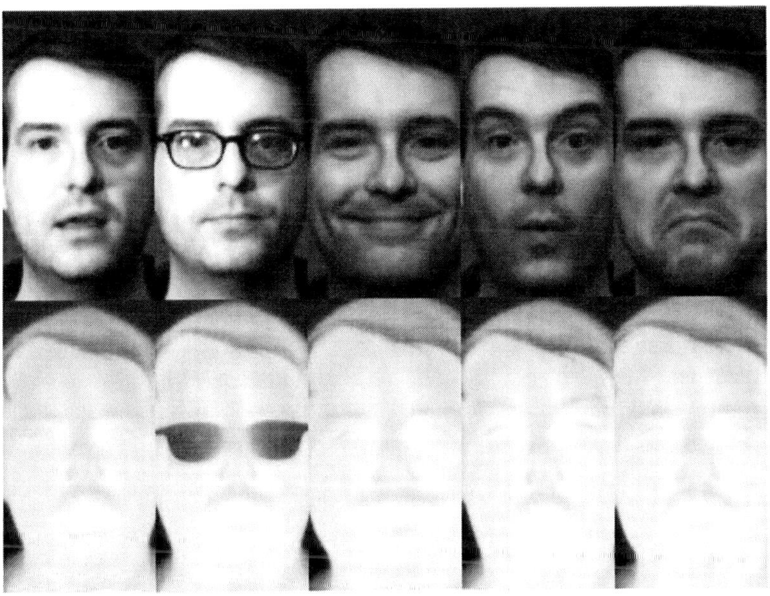

Figure 2. Equinox database examples.

4.2. PUCV-Visible Thermal-Face Database Description

The PUCV-VTF database [26] was developed to contribute to the research community interested in face recognition. The visible images were acquired with a PS3Eye camera [27], with the following specifications: image resolution of 640 × 480 pixels, field of view between 56° and 75° and a frame rate of 120/60 frames per second.

Table 1. Equinox Database Description. The quantity of images the test sets is detailed in the table.

Sets	Description	Subjects	Illuminations	Image Number
VA	Vowel Frames	All Subjects	All illuminations	729 images
EA	Expressions frames	All Subjects	All illuminations	729 images
VF	Vowel frames	All Subjects	Frontal illumination	243 images
EF	Expressions frames	All Subjects	Frontal illumination	243 images
VL	Vowel frames	All Subjects	Lateral illumination	486 images
EL	Expressions frames	All Subjects	Lateral illumination	486 images
VG	Vowel frames	Subjects Using Glasses	All illuminations	324 images
EG	Expressions frames	Subjects Using Glasses	All illuminations	324 images
RR	Random 500 frames	Chosen at random	All illuminations	500 images

Thermal images were acquired with a FLIR Tau 2 camera [28], with the following specifications: image resolution of 640 × 512 pixels and 14 bits, field of view between 56° and 69°, frame rate of 30 frames per second, thermal sensibility of 50 mK and a spectrum range between 7.5 and 13.5 µm.

The physical setup is shown in Figure 3. This consists of a tripod where the two sensors (FLIR Tau 2 and PS3 Eye) are positioned with a separation of 6.5 cm. These cameras are at a height of 115 cm from the floor and at a distance of 91.5 cm from the background. The background is a white wall, which was chosen to avoid thermal and visible changes that might affect the acquisition of the images.

The database includes 76 individuals with five subsets, in a total of 12,160 images in both visible and thermal spectra. The sets comprise normal faces (frontal condition), frown, glasses, smile and vowel. Only one image per subject is used to create the sets. The normal set was considered as the gallery, as it is neutral, while the rest were used as test sets. The characteristics of the sets are the following: normal faces indicate a serious pose of the individual; frown indicates anger; glasses are images of the individual using transparent or dark glasses at random; smile shows some grinning; and vowel faces are images of the individuals articulating a vowel of their choice. We use the following notations for each set: Nml for normal, Frn for frown, Gls for glasses, Slm for smile and Vwl for vowel (see the examples in Figure 4).

Figure 3. Physical setup of the PUCV- VisibleThermal-Face (VTF) database acquisition system.

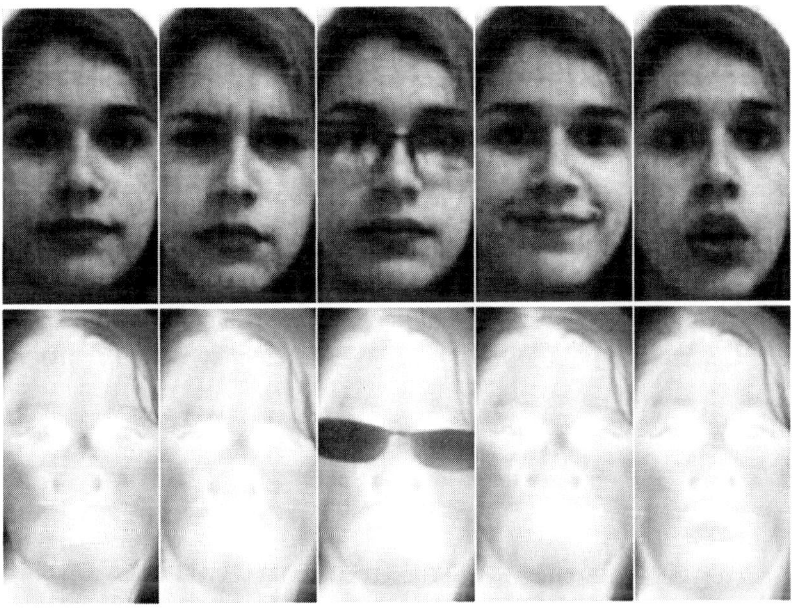

Figure 4. PUCV-VTF database examples.

All of the images were cropped and aligned according to the coordinates of the eyes, which were marked manually. The length and proportion of the distance between the eyes and the borders were made in line with the Equinox database (81 × 150 pixels with 42 pixels between the eyes). The visible and thermal images were normalized using the following Equation (2), which consists of applying a linear mapping to the pixels intensity values in the range [N_{min}, N_{max}]:

$$I_{norm}(i,j)=\frac{I(i,j)-I_{\min}}{I_{\max}-I_{\min}}(N_{max}-N_{min})+N_{min}\ \forall(i,j)\in\Omega \quad (2)$$

where I_{min} and I_{max} are the minimum and maximum values in the image:$I_{min}=\min_{(i,j)\in\Omega} I(i,j)$ and $I_{max}=\max_{(i,j)\in\Omega} I(i,j)$. In the experiments, the value range of [0,255] is used.

5. EXPERIMENTS

In order to evaluate the proposed fusion methodology using GA, the face recognition experiments were performed using two databases: Equinox and PUCV-VTF. The first experiment consists of comparing face recognition results from the fusion method against other face recognition methods and the fusion results obtained from [18,19,20], thus obtaining the best configuration of our methodology and the optimal population to achieve high recognition rates. The GA was run in order to obtain the best genetic codes for the fusion method. After the experiment, the optimal population was selected. This is then validated with the PUCV-VTF database, comparing the results with the visible and thermal algorithms used in the literature.

5.1. Experiment 1: Optimal Population in the Equinox Database

The main idea of this experiment is to obtain the best genetic codes with the GA process for the proposed fusion method described in Section 3. The fusion methods used 2 variants composed of the combination of visible and thermal descriptors: LDP-LBP and LBP-LBP descriptors. The fusion method using the LDP3 operator for visible spectrum and LBP descriptor in the thermal is named FD-LDP-LBP, and FD-LBP-LBP is the fusion method using the LBP descriptor for both spectra.

To achieve the best genetic codes, only two subsets (the most representative) of the Equinox database are used to obtain the optimal weights. In the experiment, the GA is trained using the EA set as the gallery set and the RR set as the test set, since this set is random and represents more variability in the images (lighting conditions, expression, artefacts). The GA was applied to the gallery and test sets, finding the best possible population using the proposed fusion methodology following the steps shown in Section 3. The evolution of the genetic algorithm (fitness value) is shown in Figure 5 for both FD-LDP-LBP and FD-LBP-LBP. The optimal final population represented by 100 different individuals was found, achieving the same recognition rate for the gallery and test sets, reaching almost 98% for both fusion descriptor methods. Note that FD-LBP-LBP showed a faster convergence than FD-LDP-LBP to achieve the optimum.

Once the GA process had been performed, the optimal population was tested using the entire Equinox database. In addition, the face recognition rate obtained using the fusion method was compared with the results obtained using the face recognition methods shown in Section 3 and the fusion results obtained from KPCA (kernel PCA) and KFLD (kernel Fisher linear discriminant) [19], wavelets and PCA [18] and discrete wavelet transform (DWT) [20]. The recognition experiments were performed using the face recognition methods applied to the gallery and test sets for the Equinox database. For each gallery and test set, the top one recognition was computed. This consists of obtaining the most similar image as a correct rate from the gallery/test sets. In addition, the average recognition rate of each result from gallery/test pairs of the entire Equinox database was used as the evaluation criteria. This is calculated as the mean of the recognition rate for each set. The results of the fusion approaches obtained from the literature [18,19,20] do not include the same gallery/test sets used in this work, because those studies often removed the case of the subject with glasses. In order to obtain a fair evaluation, the best results reported in each were considered to obtain a comparison criterion with the proposed fusion method.

The results obtained are shown in Table 2 after applying 100 genetic codes obtained from the GA process for both fusion methods. All 100 genetic codes obtained better recognition rate results in comparison with current face recognition methods, even the worst code obtained. Table 2 shows the results from the face recognition method and only two results of the fusion descriptors, *i.e.*, the best variants for FD-LDP-LBP and FD-LBP-LBP, which were chosen to show the effectiveness of the proposed fusion method. In addition, comparing all of the results of the face recognition methods, our proposed methods increase recognition performance to 96.99%. Thus, it is possible to obtain a unique genetic code that ensures the best performance for the face recognition system with the Equinox database using FD-LDP-LBP. In addition, the FD-LBP-LBP also increases performance, achieving 95.34% for recognition and surpassing the classic visible and thermal methods. The fusion methods from the literature obtained high performance rates, achieving ~96% recognition rates; however, these fusion results are less than the results obtained from the proposed fusion method. Previous fusion techniques do not consider difficult cases, such as glasses, which further enhances our results. We believe that the high recognition rates achieved by our fusion methodology are due to the local descriptors used for the face recognition, which are robust to occlusion. Furthermore, the descriptors are not affected by lighting conditions, and the genetic algorithm efficiently combines features for the face description. For these reasons, the next experiments will not consider the evaluation of the classic fusion methods. Note that without using the proposed fusion method, the best variant for thermal face recognition is LBP, obtaining 95.32%, but in the visible case, the best method is LDP3 with only an 88.20% recognition rate.

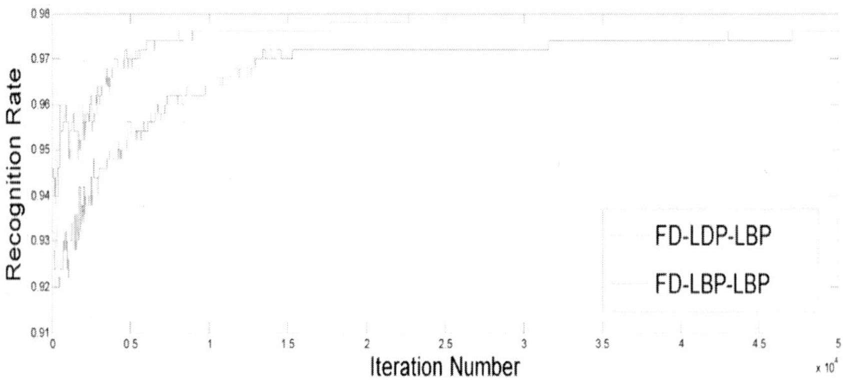

Figure 5. Fitness value evolution for the genetic algorithm. FD, fusion descriptor; LDP, local derivative pattern; LBP, local binary pattern.

Figure 6 presents an analysis of the behavior of the best variants of face recognition methods and the proposed fusion method for all Equinox galleries. The robustness of each method is evaluated by calculating the recognition rate using different conditions, such as expressions, gestures, lighting and glasses, by using different test sets from the Equinox database. Thus, for both spectra, it can be seen that FD-LDP-LBP and FD-LBP-LBP are more robust than other methods to changes in expression and the use of glasses, since they acquire visible and thermal characteristics given by the optimal weights. Additionally, considering the visible spectrum results, most of the visible methods are affected by glasses and illumination, whereas the thermal methods are more robust to expression and illumination conditions, but are affected by glasses, with LDP3 and GJD being the worst methods in this spectrum.

Table 2. Average recognition rate for the proposed fusion method against all face recognition methods for the Equinox database. Results in bold show the best variation. HI, histogram intersection; WLD, Weber law descriptor; GJD, Gabor jet descriptor; KPCA, kernel PCA; KFLD, kernel Fisher linear discriminant.

Method	Visible (%)	Thermal (%)
LBP_HI_80	81.60	**93.89**
LBP_HI_32	80.11	**95.32**
LDP2_X2_32	83.12	**88.98**
LDP3_HI_256	**88.20**	61.10
LDP3_X2_32	**85.22**	73.30
HOG_X2_256	73.59	**94.03**
WLD_HI_80	78.89	**84.86**
WLD_X2_80	38.05	**85.19**
GJD	**85.35**	70.07
FD-LDP-LBP	**96.99**	
FD-LBP-LBP	95.64	
Wavelets [18]	93.50	
PCA [18]	92.90	
KPCA [19]	82.70	
KFLD [19]	96.30	
Wavelets [20]	96.10	

To finish our experiment and as a means of representation, the best genetic codes obtained from FD-LDP-LBP were selected, showing that parts of the visible and thermal image of a human face are useful for face recognition process. For example, given its weight values, the best genetic code found shows that there are areas for face recognition that are more significant in the visible spectrum, and others are more important in the thermal spectrum. Figure 7 shows a face divided into 32 regions, where each region is representing by a color, blue or red, representing the information of the optimal population found with the GA. In addition, in Figure 7, it can be seen that the proposed method chose to give more relevance to the thermal regions (red parts: noose, lips and chin), because these are more invariant to changes in lighting and pose, but in the areas around the forehead and hair, it gives more relevance to visible areas (blue parts), as the thermal imaging has no information. In the next section, the two fusion descriptions, FD-LDP-LBP and FD-LBP-LBP, and the best population are used to analyze another database to validate the performance of our fusion methods.

Figure 6. Average recognition rates for the Equinox database. Different test sets against the entire gallery sets.

Figure 7. Optimal genetic code obtained for the GA.

5.2. Experiment 2: Fusion Method Validation Using the PUCV-VTF Database

The second experiment consist of evaluating the PUCV-VTF database with the fusion methodology *versus* the visible and thermal face recognition methods to compare and validate the proposed fusion descriptor. Firstly, the GA was not re-implemented to obtain the best population, since the best population obtained with the Equinox database was used. Secondly, the experiments were conducted using the top one criterion to obtain the recognition rate. Similar to the Equinox database, the different test sets (frown, glasses, smile and vowels) were compared with the gallery (normal) using all of the methods from Section 2 and the proposed fusion method, selecting the best results, which are shown in Table 3. From Table 3, it can be seen that the results are consistent with the Equinox database, where the fusion methodologies achieve the best results and very high recognition rates (99.01%) for FD-LBP-LBP. In addition, one fusion method, FD-LDP-LBP, and the thermal LBP variant, LBP_HI_80, obtain similar results: 98.68%. Note that thermal methods attain high performance in recognition rates with local descriptors, such as LBP, HOG and WLD.

To show the performance of each test set *versus* the gallery set, Table 4 was created. It shows the average recognition rate of all test sets (frown, glasses, smile and vowels) *versus* the gallery set (normal). We can see that for the frown set, the two proposed fusion descriptors obtain the best performance of 98.68%, where, for example, the GJD and LDP variants in the visible and thermal perform very poorly. For the glasses set, most face recognition methods decrease their recognition rates, but the FD-LBP-LBP and the thermal LBP_HI_80 obtain the best results, remaining very robust with the use of this artefact. With the smile set, all methods are very robust to this pose variation in the thermal spectrum, where LBP, WLD, GJD and the proposed fusion methods achieve 100% recognition rates. For the vowel set, something different occurs, because here, the thermal LBP obtains 100% performance and surpasses the proposed fusion descriptor. We believe the reason is that the visible features do not give more information for the proposed fusion descriptor with the vowel set. Finally, the average results of the entire gallery/test sets are compared to identify the best performance, which is seen for the fusion descriptor methodology (FD-LBP-LBP). The result of a 99.01% recognition rate validates the proposed fusion methodology, showing high results for both the Equinox and PUCV VTF databases.

For both the Equinox and PUCV-VTF databases, the fusion methodology obtains high face recognition performance, achieving the best recognition rates. These results are very encouraging, since, through the use of both spectra, it is possible to combine better features than in the visible and thermal spectra separately. Thus, it can be concluded that the proposed fusion descriptor using a genetic algorithm increases the recognition rates in a visible/thermal face recognition system.

Table 3. Average recognition rate for the proposed fusion method against all face recognition methods for the PUCV-VTF database. Results in bold show the best variation.

Variant	Visible (%)	Thermal (%)
LBP_HI_80	91.45	**98.68**
LBP_X2_32	96.71	**98.03**
LDP2_HI_32	**94.08**	91.78
LDP2_HI_256	91.12	**93.09**
LDP3_X2_32	78.95	**83.22**
LDP3_EU_256	74.01	**86.51**
HOG_EU_256	93.09	**97.04**
HOG_X2_256	92.76	**97.70**
WLD_HI_80	90.46	**97.70**
GJD	80.26	**92.76**
FD-LDP-LBP	98.68	
FD-LBP-LBP	**99.01**	

6. CONCLUSIONS

In this study, a new fusion methodology to combine visible and thermal descriptors for face recognition systems was presented. The proposed fusion methodology was evaluated using the standard visible and thermal database, Equinox, and a new database, entitled PUCV-VisibleThermal-Face, created to validate the proposed fusion method and to be used in further research on visible and thermal face recognition. In addition, this study compared the recognition rates of current face recognition methods, such as local derivative patterns, local binary patterns, Weber local descriptors, histograms of oriented gradients and Gabor jet descriptors, and classic fusion techniques used commonly in the literature; against the proposed methodology of fusion descriptors using a genetic algorithm to find the best population and, thus, to achieve an optimal face recognition rate.

Table 4. Average recognition rate for the PUCV-VTF database. Different test sets against the entire gallery sets. In bold are the best results per test set.

Variant	Average Recognition Rate-Test Sets				Average
	Frown	Glasses	Smile	Vowels	
LBP_X2_32 (Visible)	96.05	96.05	98.68	96.05	96.71
LBP_HI_80 (Thermal)	96.05	**98.68**	**100.00**	**100.00**	98.68
LDP2_HI_32 (Visible)	94.74	93.42	94.74	93.42	94.08
LDP2_HI_256 (Thermal)	89.47	92.11	97.37	93.42	93.09
LDP3_X2_32 (Visible)	65.79	77.63	86.84	85.53	78.95
LDP3_EU_256 (Thermal)	82.89	78.95	98.68	85.53	86.51
HOG_EU_256 (Visible)	88.16	92.11	96.05	96.05	93.09
HOG_X2_256 (Thermal)	96.05	97.37	98.68	98.68	93.09
WLD_HI_80 (Visible)	85.53	85.53	97.37	93.42	90.46
WLD_HI_80 (Thermal)	96.05	97.37	**100.00**	97.37	97.70
GJD (Visible)	60.53	69.74	97.37	93.42	80.26
GJD (Thermal)	92.11	82.89	**100.00**	96.05	92.76
FD-LDP-LBP	**98.68**	97.37	**100.00**	98.68	98.68
FD-LBP-LBP	**98.68**	**98.68**	**100.00**	98.68	**99.01**

To achieve the optimal population for the fusion method, the methodology was tested using a part of the Equinox database, achieving a fitness convergence of 98% recognition rate. Thus, with the optimal population, the fusion descriptor was evaluated using the entire Equinox database and the associated methodology [7,12]. The results show that the two fusion descriptors used (FD-LDP-LBP and FD-LBP-LBP) obtain high performances, surpassing the visible and thermal results of the current face recognition methods and other classic fusion approaches used in the literature. In addition, observing the results obtained by the fusion descriptor, the methodology is very robust to changes in illumination and expression, because it combines thermal and visible information, allowing it to choose optimal face areas where one spectrum is more representative than the other. However, when the individual uses an artefact, such as glasses, and although the fusion descriptor results are superior to the visible/thermal face recognition methods, the performance decreases slightly, since the optimal weights are statically obtained with images that randomly contain glasses and no glasses, as reported in [18,20]. We believe that this can be solved by increasing the variability of the images used to obtain the optimal population (more examples to increase the fitness values/recognition rates).

The fusion methodology was validated using the average recognition rate obtained from the evaluation of current visible/thermal face recognition rates and the PUCV-VTF database. The results show the high performance obtained by the fusion descriptor (FD-LBP-LBP), achieving a recognition rate of 99.01%.

This result validates the fusion method, as it shows that the fusion descriptor operates with high recognition rates and is robust to images with different variabilities, such as expressions (frown, smile), gestures and artefacts, such as glasses.

The final conclusion on the fusion descriptors is that combining visible and thermal descriptors increases the performance of the face recognition system. The analysis of the fusion descriptor and the evaluation with the two databases shows that using a genetic algorithm to optimize the recognition rates gives an optimal population and leads to the best performance, due to the selection of the optimal weights and, thus, obtaining the best possible combination of visible and thermal features. We conclude that the proposed fusion method is very robust to variations and artefacts and obtains high recognition rates. However, as a future study, these results could be increased with new techniques based on evolutionary algorithms [29], or with the combination of new face recognition descriptors [30,31], or new texture classification methods published in recent literature.

ACKNOWLEDGMENTS

This research was funded by FONDECYT-Chile Grant 11130466.

AUTHOR CONTRIBUTIONS

Gabriel Hermosilla and Gonzalo Farias conceived and designed the experiments; Francisco Gallardo performed the experiments; Cesar San Martin contributed in the data analysis; All author contributed to write the paper.

CONFLICTS OF INTEREST

The authors declare no conflict of interest.

REFERENCES

1. Akhloufi, M.; Bendada, A.; Batsale, J.C. State of the art in infrared face recognition. *Quant. InfraRed Thermogr. J.* **2008**, *5*, 3–26.

2. FLIR, The Ultimate Infrared Handbook for R&D Professionals. Available online: http://www.flirmedia.com/MMC/THG/Brochures/T559243/T559243_EN.pdf (accessed on 22 July 2015).

3. Chen, X.; Flynn, P.J.; Bowyer, K.W. PCA-based face recognition in infrared imagery: Baseline and comparative studies. In Proceedings of the IEEE International Workshop on Analysis and Modeling of Faces and Gestures (AMFG 2003), Nice, France, 17 October 2003; pp. 127–134.

4. Socolinsky, D.; Selinger, A. Thermal Face Recognition over Time. In Proceedings of the 17th International Conference on Pattern Recognition (ICPR'04), 23–26 August 2004; Volume 4, pp. 187–190.

5. Holland, J.H. *Adaptation in Natural and Artificial Systems*; MIT Press: Cambridge, MA, USA, 1992.

6. Sastry, K.; Goldberg, D.E.; Kendall, G. *Search Methodologies: Introductory Tutorials in Optimization, Search and Decision Support Techniques*; Burke, E., Kendall, G., Eds.; Springer: Berlin, Germany, 2005.

7. Hermosilla, G.; Ruiz-del-Solar, J.; Verschae, R.; Correa, M. A Comparative Study of Thermal Face Recognition Methods in Unconstrained Environments. *Pattern Recognit.* **2012**, *45*, 2445–2459.

8. Ruiz-del-Solar, J.; Verschae, R.; Correa, M. Recognition of Faces in Unconstrained Environments: A Comparative Study. *EURASIP J. Adv. Signal Process.* **2009**, *2009*, 1–19.

9. Selinger, A.; Socolinsky, D. Appearance-Based Facial Recognition Using Visible and Thermal Imagery: A Comparative Study. Technical Report; Equinox Corporation: New York, NY, USA, 2001.

10. Socolinsky, D.; Selinger, A. A comparative analysis of face recognition performance with visible and thermal infrared imagery. In Proceedings of the 16th International Conference on Pattern Recognition (ICPR), Quebec, QC, Canada, 11–15 August 2002; pp. 217–222.

11. Zou, J.; Ji, Q.; Nagy, G. A Comparative Study of Local Matching Approach for Face Recognition. *IEEE Trans. Image Process.* **2007**, *16*, 2617–2628.

12. Hermosilla, G.; Farias, G.; Vargas, H.; Gallardo, F.; San-Martin, C. Thermal Face Recognition Using Local Patterns. In *Progress in Pattern Recognition, Image Analysis, Computer Vision, and Applications*; Lecture Notes in Computer Science; Springer: Puerto Vallarta, Jalisco, Mexico, 2014; Volume 8827, pp. 486–497.

13. Ahonen, T.; Hadid, A.; Pietikainen, M. Face Description with Local Binary Patterns: Application to Face Recognition. *IEEE Trans. Pattern Anal. Mach. Intell.* **2006**, *28*, 2037–2041.

14. Chen, J.; Shan, S.; He, C.; Zhao, G.; Pietikäinen, M.; Chen, C.X.; Gao, W. WLD: A Robust Local Image Descriptor. *IEEE Trans. Pattern Anal. Mach. Intell.* **2010**, *32*, 1705–1720.

15. Zhang, B.; Gao, Y.; Zhao, S.; Liu, J. Local Derivative Pattern Versus Local Binary Pattern: Face Recognition with High-Order Local Pattern Descriptor. *IEEE Trans. Image Process.* **2010**, *19*, 533–544.

16. Dalal, N.; Triggs, B. Histograms of Oriented Gradients for Human Detection. In Proceedings of the 2005 IEEE Computer Society Conference on Computer Vision and Pattern Recognition (CVPR'05), San Diego, CA, USA, 25 June 2005; Volume 1, pp. 886–893.

17. Deniz, O.; Bueno, G.; Salido, J.; de la Torre, F. Face recognition using histograms of oriented gradients. *Pattern Recognit. Lett.* **2011**, *32*, 1598–1603.

18. Bebis, G.; Gyaourova, A.; Singh, S.; Pavlidis, I. Face recognition by fusing thermal infrared and visible imagery. *Image Vis. Comput.* **2006**, *24*, 727–742.

19. Desa, S.; Hati, S. IR and Visible Face Recognition using Fusion of Kernel Based Features. In Proceedings of the 19th International Conference on Pattern Recognition (ICPR 2008), Tampa, FL, USA, 8–11 December 2008; pp. 1–4.

20. Kwon, O.K.; Kong, S.G. Multiscale fusion of visual and thermal images for robust face recognition. In Proceedings of the IEEE International Conference Computational Intelligence for Homeland Security and Personal Safety, Orlando, FL, USA, 31 March–1 April 2005; Volume IV, pp. 112–116.

21. Abidi, B.; Huq, S.; Abidi, M. Fusion of visual, thermal, and range as a solution to illumination and pose restrictions in face recognition. In Proceedings of the IEEE International Carnahan Conference on Security Technology, Albuquerque, NM, USA, 11–14 October 2004; pp. 325–330.

22. Pop, F.M.; Gordan, M.; Florea, C.; Vlaicu, A. Fusion based approach for thermal and visible face recognition under pose and expresivity variation. In Proceedings of the Roedunet International Conference (RoEduNet), Sibiu, Romania, 24–26 June 2010; pp. 61–66.

23. Erkanli, S.; Li, J.; Oguslu, E. Fusion of Visual and Thermal Images Using Genetic Algorithms. In *Bio-Inspired Computational Algorithms and Their Applications*; Gao, S., Ed.; InTech: Rijeka, Croatia, 2012.

24. Equinox Corporation. Available online: http://www.equinoxsensors.com/products/HID.html(accessed on 1 March 2011).

25. Socolinsky, D.; Wolff, L.; Neuheisel, J.; Eveland, C. Illumination invariant face recognition using thermal infrared imagery. In Proceedings of the IEEE Conference on Computer Vision and Pattern Recognition, Kauai, HI, USA, 8–14 December, 2001; Volume 1, pp. I-527–I-534.

26. PUCV-VTF database. Available online: http://goo.gl/5bWnYp (accessed on 22 July 2015).

27. PS3 eye camera. Available online: http://us.playstation.com/ps3/accessories/playstation-eye-camera-ps3.html (accessed on 5 March 2015).

28. IR Camera FLIR TAU 2. Available online: http:// www.flir. com/cvs/cores/uncooled/products/tau (accessed on 22 July 2015).

29. Jothi, J.A.A.; Rajam, V.M.A. Segmentation of Nuclei from Breast Histopathology Images Using PSO-based Otsu's Multilevel Thresholding. In *Artificial Intelligence and Evolutionary Algorithms in Engineering Systems*; Advances in Intelligent Systems and Computing; Springer: Kumaracoil, India, 2015; Volume 325, pp. 835–843.

30. Liu, H.; Hsu, S.; Huang, C. Single-sample-per-person-based face recognition using fast Discriminative Multi-manifold Analysis. In Proceedings of the Asia-Pacific Signal and Information Processing Association, 2014 Annual Summit and Conference (APSIPA), Siem Reap, Cambodia, 9–12 December 2014; pp. 1–9.

31. Hu, S.; Choi, J., Chan, A.; Schwartz, W. Thermal-to-visible face recognition using partial least squares. *J. Opt. Soc. Am.* **2015**, *32*, 431–442.

Chapter 2

Face recognition using color local binary pattern from mutually independent color channels

Gholamreza Anbarjafari

Department of Electrical and Electronic Engineering, Cyprus International

University, Lefkoşa, KKTC, Mersin 10, Turkey

ABSTRACT

In this article, a high performance face recognition system based on local binary pattern (LBP) using the probability distribution functions (PDFs) of pixels in different mutually independent color channels which are robust to frontal homogenous illumination and planer rotation is proposed. The illumination of faces is enhanced by using the state-of-the-art technique which is using discrete wavelet transform and singular value decomposition. After equalization, face images are segmented by using local successive mean quantization transform followed by skin color-based face detection system. Kullback–Leibler distance between the concatenated PDFs of a given face obtained by LBP and the concatenated PDFs of each face in the database is used as a metric in the recognition process. Various decision fusion techniques have been used in order to improve the recognition rate. The proposed system has been tested on the FERET, HP, and Bosphorus face databases. The proposed system is compared with conventional and the state-of-the-art techniques. The recognition rates obtained using FVF approach for FERET database is 99.78% compared with 79.60 and 68.80% for conventional gray-scale LBP and principle component analysis-based face recognition techniques, respectively.

KEYWORDS

Illumination robust, Local binary pattern, Face recognition, Probability distribution function, Discrete wavelet transform, Kullback–Leibler distance

1. INTRODUCTION

Face recognition has been one of the most interesting research topics for over the past half century. During this period, many methods such as principle component analysis (PCA), linear discriminant analysis (LDA), independent component analysis (ICA), etc., have been introduced [1–5]. Many of these methods are based on gray-scale images; however, color images are increasingly being used since they add additional biometric information for face recognition [6–8]. As reported by Demirel and Anbarjafari [6, 8], color probability distribution functions (PDFs) of a face image can be considered as the signature of the face, which can be used to represent the face image in a low-dimensional space. It is known that PDF of an image is a normalized version of an image histogram [9]. PDF recently has been used in many applications of image processing such as object detection, face localization, and face recognition [6, 8–12].

One of the most important steps in a face recognition system is face segmentation. There are various methods for segmentation of the faces such as skin color-based face segmentation [13, 14], Viola–Jones [15] face detection system, local successive mean quantization transform (SMQT)-based face detection [16, 17]. In this study, we are using local SMQT-based face segmentation followed by skin color-based face segmentation. This procedure will reduce the effect of background on the rectangle-shape segmented face image.

In this article, the PDF-based face recognition will be studied analytically and then LBP will be used in order to boost the recognition performance. Also in this article, instead of experimentally choosing PDFs of *HSI* and *YCbCr* color channels [6, 8], analytically specific color channels have been selected. Furthermore, analytical studies of false acceptance rate (FAR) and false rejection rate (FRR) analysis are included in the third section. The head pose (HP) face database [18] with 15 subjects, a subset of 50 subjects from the

FERET [19] database with faces containing varying poses changing from –90° to +90° of rotation around the vertical axis passing through the neck (the same subset as Demirel and Anbarjafari used in [6, 8]), and Bosphorus face database [20] with 105 subjects with varying frontal illuminations, poses, expressions, and occlusions were used to test the proposed system.

2. FACIAL IMAGES PRE-PROCESSING

2.1 Image illumination enhancement

In many image processing applications, the general histogram equalization (GHE) is one of the simplest and most effective primitives for contrast enhancement [21], which attempts to produce an output histogram that is uniform [22]. One of the disadvantages of the GHE is that the information laid on the histogram or PDF of the image will be lost. Demirel and Anbarjafari [6] have showed that the PDF of face images can be used for face recognition, hence preserving the shape of PDF of the image is of vital importance. Therefore, GHE is not a suitable technique for illumination enhancement of face images. Also, it is known that GHE often produces unrealistic effects in images. After the introduction of GHE, researchers came out with better technique which deals with equalization of portion of the image at a time, called local histogram equalization (LHE). LHE can be expressed as follows: GHE can be applied independently to small regions of the image. Most small regions will be very self-similar. If the image is made up of discrete regions, most small regions will lie entirely within one or the other region. If the image has more gradual large-scale variation, most small regions will contain only a small portion of the large-scale variation.

However, the contrast issue is yet to be improved and even these days many researchers are proposing new techniques for image equalization. DHE is obtained from dynamic histogram specification [23] which generates the specified histogram dynamically from the input image.

Demirel and Anbarjafari [6, 24] developed singular value decomposition (SVD)-based image equalization (SVE) technique, which is based on equalizing the singular value (SV) matrix obtained by SVD. Since an image can be considered as a numeric matrix, SVD of an image, A, can be written as follows

$$A = U_A \Sigma_A V_A^T \tag{1}$$

where U_A and V_A are orthogonal square matrices known as hanger and aligner, respectively, and Σ_A matrix contains the sorted SVs on its main diagonal. The idea of using SVD for image equalization comes from this fact that Σ_A contains the intensity information of the given image [25]. The objective of SVE [24] is to equalize a low-contrast image in such a way that the mean moves towards the neighborhood of 8-bit mean gray value 128 in the way that the general pattern of the PDF of the image is preserved. Demirel and Anbarjafari [6] used SVD to deal with the illumination problem in their proposed face recognition system. SVE can be described in the following way: the ratio of the largest SV of the generated normalized matrix over a normalized image. This coefficient can be used to regenerate an equalized image. This task is eliminating the illumination problem. It is important to mention that techniques such as DHE or SVE are preserving the general pattern of the PDF of an image.

The proposed method is robust to the frontal homogenous illumination changes and this robustness is achieved by using a state-of-the-art technique which is based on discrete wavelet transform (DWT) and SVD [26]. The two-dimensional wavelet decomposition of an image is performed by applying the one-dimensional DWT along the rows of the image first, and then the results are decomposed along the columns. This operation results in four decomposed sub-band images refer to low-low (LL), low-high (LH), high-low (HL), and high-high (HH). The frequency components of those sub-band images cover the frequency components of the original image. DWT is used to separate the input image into different frequency sub-bands, where LL sub-band concentrates the illumination information. The method is benefiting from the fact that an intensity image (an 8-bit intensity image in a given color channel), A, can be decomposed into multiplication of three matrices, by using SVD. The first SV has the highest impact on the illumination of the image, hence updating this SV will directly effect on the illumination of A. Also it is known that the equalized image by using GHE will result into a visually good looking image [27]. Furthermore, in wavelet domain the illumination information is laid in the LL sub-band. By knowing this information, the method is modifying the SVs by using a correction coefficient, ζ:

$$\zeta = \frac{\max\left(\Sigma_{LL\hat{A}}\right)}{\max\left(\Sigma_{LL_A}\right)} \tag{2}$$

Then, the equalized image is being reconstructed by using the following equation:

$$\begin{aligned} \overline{\Sigma}_{LL_A} &= \zeta\, \Sigma_{LL_A} \\ \overline{LL_A} &= U_{LL_A} \overline{\Sigma}_{LL_A} V_{LL_A} \\ \overline{A} &= IDWT\left(\overline{LL}_A, LH_A, HL_A, HH_A\right) \end{aligned} \tag{3}$$

SVD is a computationally complex operation. As it is shown in Equation (2) only the highest (the first) SV is used. Also from elementary algebra, it is known that for a matrix, the highest SV is obtained by calculating its norm. Thus, ζ can be calculated by

$$\zeta = \frac{\|LL_{\hat{A}}\|}{\|LL_A\|} \tag{4}$$

Then the equalized image is being reconstructed by using the following equation

$$\begin{aligned} \overline{LL_A} &= \zeta LL_A \\ \overline{A} &= IDWT\left(\overline{LL}_A, LH_A, HL_A, HH_A\right) \end{aligned} \tag{5}$$

Figure 1 shows the block diagram of the proposed technique. One can implement the algorithm in different color channel in order to enhance the illumination of the color images.

Figure 2 shows the visual effect of the DWT-based illumination enhancement on a face image with three different illuminations from OULU face database [28].

2.2 Face localization and segmentation

A face is naturally recognizable by a human regardless of its many point of variation such as skin tone, facial hair, etc. Face detection is a required first step in face recognition systems [16, 29]. The most straight forward variety of face localization is the detection of a single face at a known scale and orientation, which is yet a non-trivial problem. Efficient fast face detection is an impressive goal, which is subject to face tracking that required no knowledge of previous frames [30]. Another reason that face detection is an important research issue is its role as a challenging case of a more general problem, object detection.

Skin is a widely used feature in human image processing with applications ranging from face detection [13] and person tracking [31] to content filtering [32]. Human skin can be detected by identifying the presence of skin color pixels. Many methods have been proposed for achieving this. Chai and Ngan [14] modeled the skin color in *YCbCr*color space. In their technique, pixels are classified into "skin" and "non-skin" by using four threshold values, which form a rectangular region in *CbCr* space.

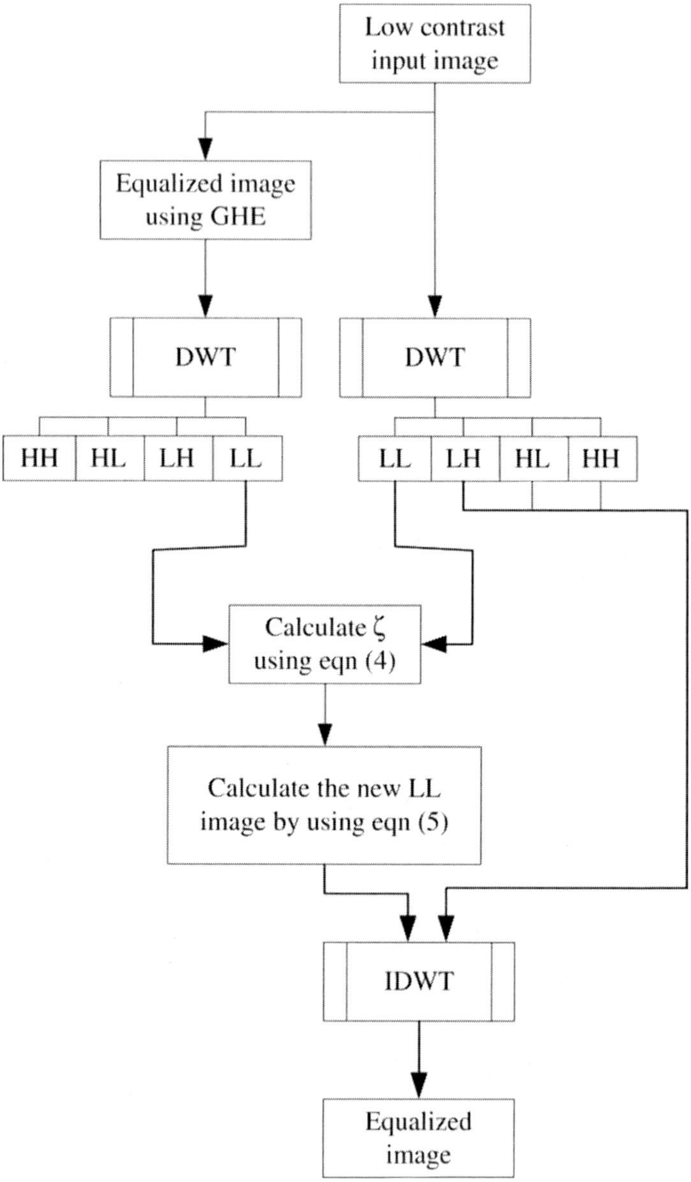

Figure 1. The detailed steps of the image equalization technique.

All color spaces such as *RGB*, *HIS*, and *YCbCr* can be used for face recognition [33]. The advantage of using *HSI*color space is its independence of knowledge of the exact percentage of red, green, or blue. Many applications such as machine vision use *HSI* color space in identifying the color of different objects.

Kjeldson and Kender [34] stated a color preference in *HSI* color space to distinguish skin regions from other segments. Skin color classification in *HSI* color space is based on hue and saturation values.

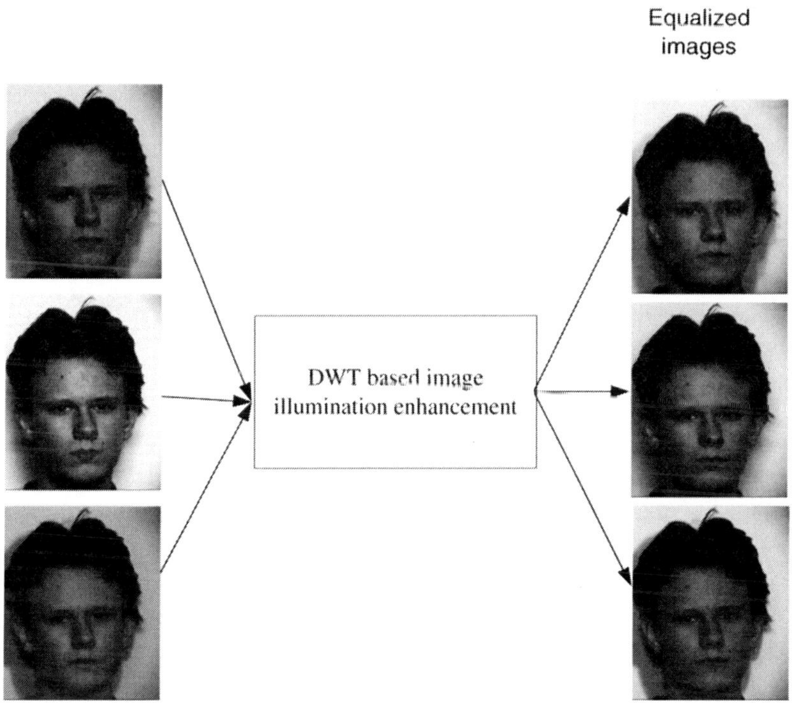

Figure 2. Sample images with different illumination from Oulu face database after and before illumination enhancement.

In [35], the threshold for hue and saturation has been modified from previous work [34] by using 2,500 face samples taken from the FERET dataset, the HP face database, and the Essex University face database. According to these samples, the thresholds for hue and saturation color channels are updated to satisfy the following threshold.

$$(H < 0.17 \quad OR \quad H > 0.63 \quad AND \quad S > 0.1 \qquad (6)$$

In order to eliminate the illumination effect from the input images, the intensity component of an image in *HSI* color space has been equalized.

A different approach to separating faces and non-faces in image space is proposed by Osuna et al. [36] and later developed and modified by Romdhani et al. [37]. Both are based on support vector machines (SVM) [38]. The key to the SVM model is the choice of a manifold that separates the face set from the non-face set. Romdhani et al. had chosen a hyperplane which maximizes minimum distance on either side. Romdhani et al. worked further on reducing the vector set in order to improve performance.

In this article, the proposed face recognition system uses local SMQT-based face detection followed by skin tone-based face localization. The SMQT can be considered as an adjustable trade-off between the number of quantization levels in the result and the computational load [39]. Local is defined to be the division of an image into blocks with a predefined size. Let x be a pixel of local D, and let's have the SMQT transform as follows

$$\mathrm{SMQT}_L : D(x) \rightarrow M(x) \tag{7}$$

where $M(x)$ is a new set of values which are insensitive to gain and bias [39]. These two properties are desired for the formation of the intensity image which is a product of reflection and illumination. A common approach to separate the reflection and illumination is based on this assumption that illumination is spatially smooth so that it can be taken as a constant in a local area. Therefore, each local pattern with similar structure will yield the similar SMQT features for a specified level, L. The spare network of winnows (SNoW) learning architecture is also employed in order to create a look-up table for classification [40].

As Nilsson et al. [39] proposed, in order to scan an image for faces, a patch of 32×32 pixels is used and also the image is downscaled and resized with a scale factor to enable the detection of faces with different sizes. The choice of the local area and the level of the SMQT are vital for successful practical operation. The level of the transform is also important in order to control the information gained from each feature. As reported in [39], the 3×3 local area and level $L = 1$ are used to be a proper balance for the classifier. The face and non-face tables are trained in order to create the split up SNoW classifier. Overlapped detections are disregarded using geometrical locations and classification score. Hence, given two detections overlapping each other, the detection with the highest classification score is kept and the other one is removed. This operation is repeated until no overlapping detection is found.

The block diagram of the local SMQT-based face detection is shown in Figure 3.

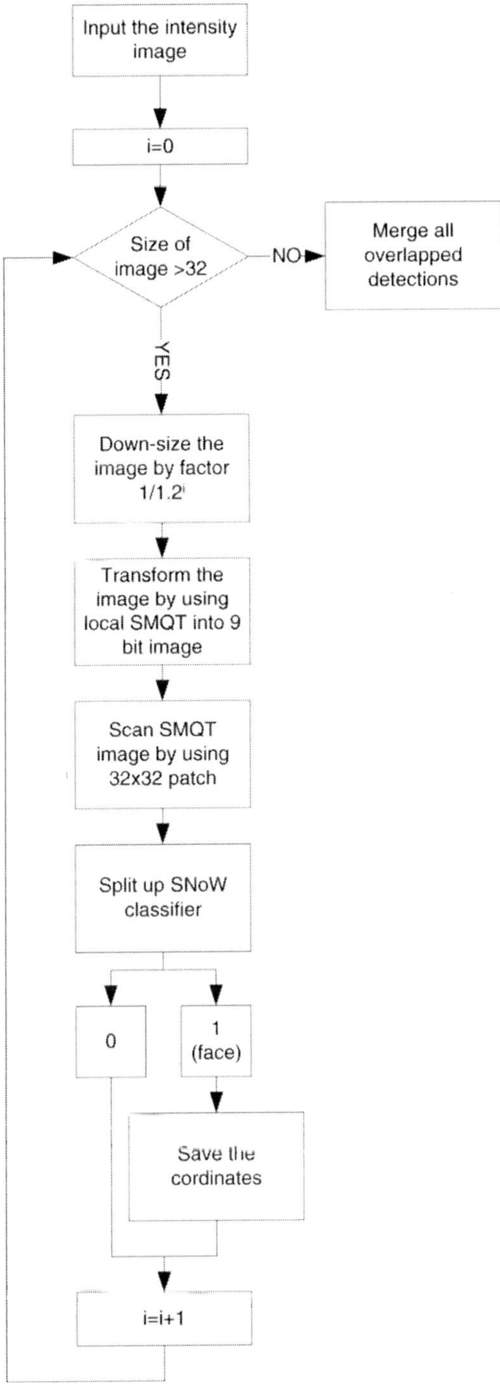

Figure 3. General steps of local SMQT-based face detection **[16]** .

Figure 4 shows the images of the three faces from Oulu face database enhanced by using DWT + SVE illumination enhancement and the segmented faces by using local SMQT.

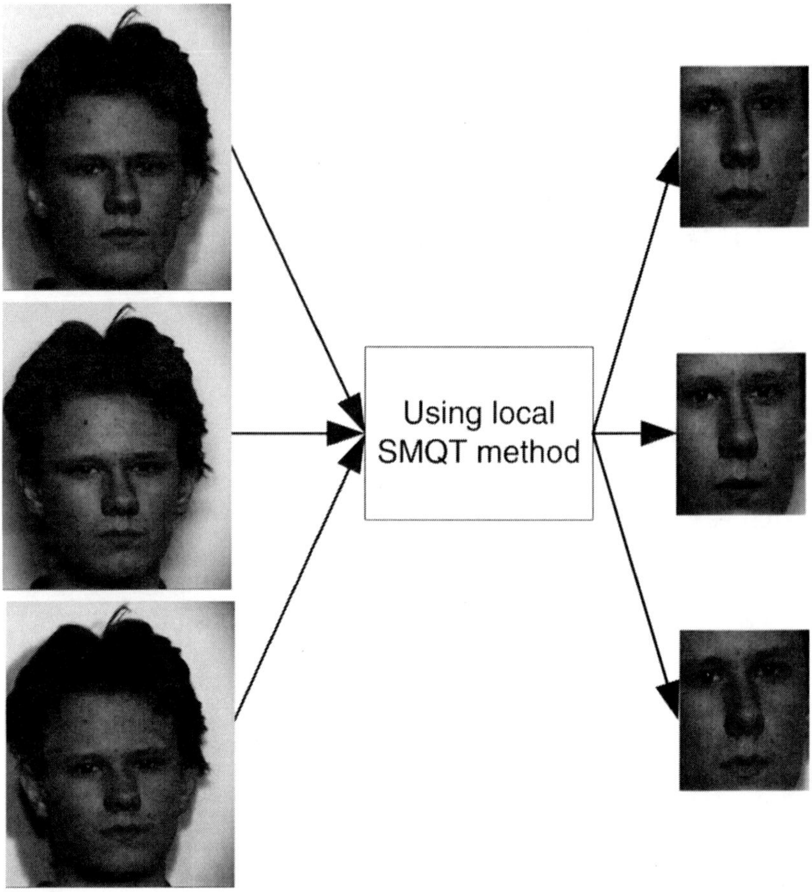

Figure 4. Face detection by using local SMQT for three samples from Oulu face database.

The local SMQT-based face detector software has several advantages which promote us in order to use it in this study, which are

- it is fast and very accurate;
- it is a state-of-the-art technique;

- if the input image does not have a face image, there will be no output, therefore in the proposed face recognition system there will be no issue of having a noise image, whose PDF is the same as a face image, as an input.

The output of the local SMQT-based face recognition is that the face image is being cropped in a rectangle frame; hence, some part of background or hair is also included in the frame of the segmented face image. In order to reduce this effect, the output of the local SMQT face detector [17] is passed through a skin color-based face segmentation system [14]. The proposed combination of the aforementioned face segmentation techniques gives an outline of Figure 5, in which a face image from the FERET face database is entered into the system.

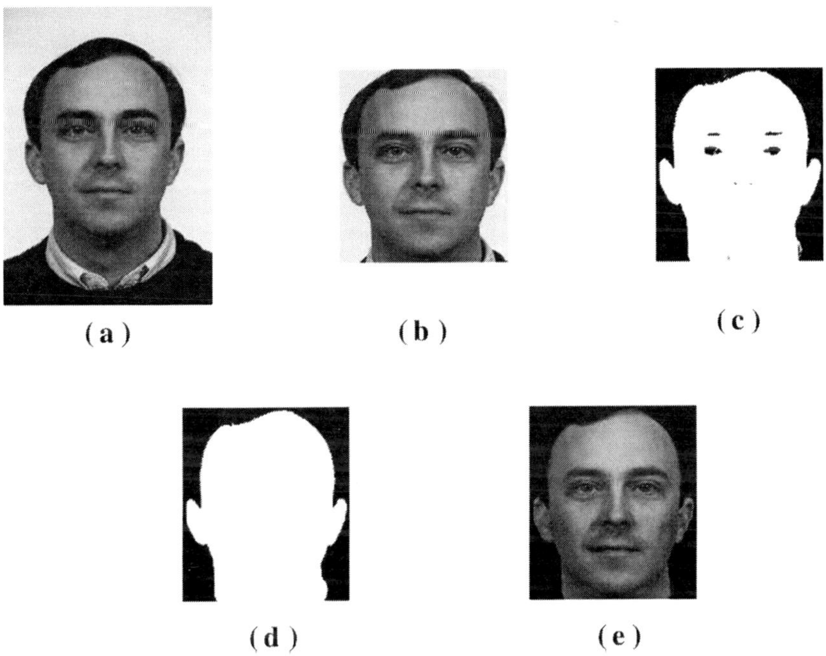

Figure 5. face localization and segmentation (a) A face image from FERET face database, (b) the segmented image by using local SMQT, (c) the mask generated by applying the skin color-based face detection**[14]**, (d) the mask obtained by hole filling of the previous mask, and (e) the final segmented face image.

2.3 PDF based face recognition by using LBP

Analytical point of view

In a general mathematical sense, an image PDF is simply a mapping η_i representing the probability of the pixel intensity levels that fall into various disjoint intervals, known as bins. The bin size determines the size of the PDF vector. Given a monochrome image, PDF η_j meets the following conditions, where *N* is the total number of pixels in an image and the bin size is 256:

$$N = \sum_{j=0}^{255} \eta_j \tag{8}$$

Then, PDF feature vector, *H*, is defined by

$$H = [p_0, p_1, \ldots, p_{255}]$$
$$p_\iota = \frac{\eta_\iota}{N}, \quad \iota = 0, \ldots, 255 \tag{9}$$

There are two issues in this point. First one is which metric is more suitable for classification of the PDFs of the faces. For this purpose, L1, L2, cross correlation, and Kullback–Leibler distance (KLD) have been used for the classification of gray-scale faces in the HP face database. The recognition rate is reported in Table 1. It shows that KLD performs better than the other metrics. Hence in this study, similar to Demirel and Anbarjafari's study [6, 8], we use KLD as a metric for finding the divergence between the PDFs.

Table 1. Recognition rate performance (%) of the PDF-based face recognition system for gray-scale PDFs of HP face database with 15 subjects and 10 samples per each subject obtained by using four different metrics.

# of Training Images	L1	L2	Cross Correlation	KLD
1	48.63	36.48	50.81	54.81
2	57.92	48.5	69.54	78.00
3	70.71	61.76	77.6	84.10
4	77.78	71.78	82.26	88.00
5	76.73	67.8	87.78	93.87

The second issue is the number of bin size. What will happen if the bin size drops from 256 into the smaller value? In line of this issue, the bin sizes have been changed and the recognition rate for the gray-scale face images in the HP face database has been obtained. Table 2 reports this result which shows the performance drops as the bin size decreases, due to loss of information laid on the pattern of the PDF.

Table 2. Recognition of face images of the HP face database by using different bin numbers.

Number of bins	Recognition rates (%)
256	96.00
128	93.33
64	90.67
32	82.67

One of the missing studies in Demirel and Anbarjafari's work was analysis of the discrimination power of the PDF. In order to show the discrimination power of the PDFs by using KLD, the average KLD value within and between-class distances of various databases in different color spaces are given in Table 3.

In Table 3 class discrimination, $\o_c$, is defined to be the ratio of the average between-class distance and the average within class distance which is indicating the discrimination power of different color channels. Class discrimination values show that KLD provides enough separation between classes in different color channels in PDF-based face recognition.

As reported in [6, 8], the PDF-based face recognition system can be implemented in various color channels such as *HSI* and *YCbCr* color spaces in which the luminance and chrominance are separated from each other. These multi decisions can be combined later by using various fusion techniques. But as some of these color channels are highly correlated with each other (they have high mutual entropy), there is no need to include the repeated information in decision making. Table 4 shows the mutual entropy between the channels of *HSI* color space and the *YCbCr* color space.

Table 4 shows the average mutual information in percentage between the various color channels in *HSI* and *YCbCr* color spaces for Bosphorus face databases where there exist over 4,500 face images. The high correlation between *I-Y*, *I-Cb*, and *I-Cr* color channels shows that instead of using both color spaces, using only *HSI* will have enough information in order to get conclusive recognition rate after the fusion. Also Table 4 indicates that the color channels in *YCbCr* are highly correlated with each other.

Table 3. The discrimination of the PDF by using KLD in different color channels and different databases.

DATA BASE		AVERAGE WITHIN CLASS DISTANCE	AVERAGE BETWEEN CLASS DISTANCE	CLASS DISCRIMINATION $\emptyset_c$
FERET	H	0.0542	0.3316	6.12
	S	0.0909	0.4008	4.41
	I	0.0716	0.2396	3.35
	Y	0.0260	0.1300	5.00
	Cb	0.0435	0.1689	3.88
	Cr	0.0364	0.1335	3.67
HP	H	0.0159	0.0818	5.14
	S	0.0097	0.0634	6.54
	I	0.0141	0.0612	4.34
	Y	0.0141	0.0611	4.33
	Cb	0.0132	0.0582	4.41
	Cr	0.0126	0.0573	4.55
BOS	H	0.0083	0.2166	26.11
	S	0.0093	0.9850	106.38
	I	0.0334	0.2793	8.36
	Y	0.0336	0.3385	10.07
	Cb	0.0209	0.6007	28.74
	Cr	0.0243	0.5027	20.70

Table 4. The correlation between ***HSI*** and ***YCbCr*** color channels in percentage.

	H	S	I	Y	Cb	Cr
H	100	26.49	9.36	5.28	9.29	16.13
S	26.49	100	25.97	32.91	52.51	65.58
I	9.36	25.97	100	99.05	98.32	93.39
Y	5.28	32.91	99.05	100	96.02	84.63
Cb	9.29	52.51	98.32	96.02	100	92.50
Cr	16.13	65.58	93.39	84.63	92.50	100

2.4 Local binary pattern

The local binary pattern (LBP) is a non-parametric operator which describes the local spatial structure of an image [12, 41]. Ojala et al. [41] introduced this operator and showed its high discriminative power for texture classification. At a given pixel position $(x\ y)$, LBP is defined as an ordered set of binary comparisons of pixel intensities between the center pixel and its eight neighbor pixels, as shown in Figure 6.

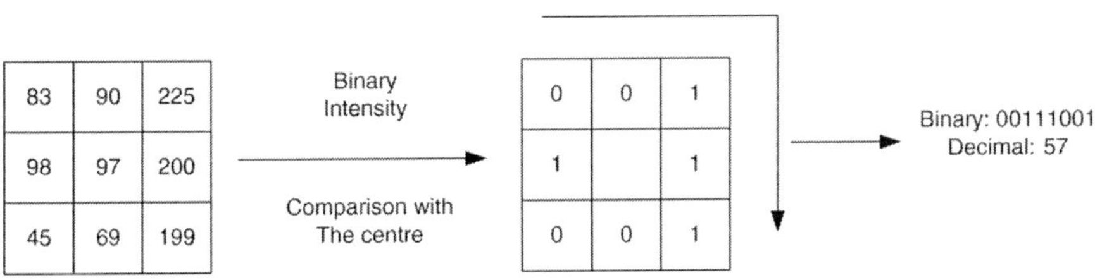

Figure 6. LBP operator.

The decimal form of the resulting 8-bit word of LBP code can be expressed as follows.

$$\mathrm{LBP}(x, y) = \sum_{n=0}^{7} 2^n s\left(i_n - i_{(x,y)}\right) \quad (10)$$

where $i_{(x,y)}$ corresponds to the gray value of the center pixel (x,y), i_n to the gray values of the eight neighbor pixels, and function $s(x)$ is defined as

$$s(x) = \begin{matrix} 1 \; if \; x \geq 0 \\ 0 \; if \; x < 0 \end{matrix} \tag{11}$$

By definition, the LBP operator is unaffected by any monotonic gray-scale transformation which preserves the pixel intensity order in a local neighborhood. Note that each bit of the LBP code has the same significance level and that two successive bit values may have a totally different meaning. Sometimes, the LBP code is referred as a kernel structure index.

Ojala et al. [42] extended their previous study to a circular neighborhood of different radius size. They used $LBP_{P,R}$notation which refers to P equally spaced pixels on a circle of radius R. Two of the main motivations of using LBP are its low computational complexity and its texture discriminative property. LBP has been used in many image processing applications such as motion detection [43], visual inspection [44], image retrieval [45], face detection [46], and face recognition [47, 48].

In most aforementioned applications, a face image was usually divided into small regions. For each region, a cumulative histogram of LBP code computed at each pixel location within the region was used as a feature vector.

Ahnon et al. [48] used LBP operator for face recognition. Their face recognition system can be explained as follows: a histogram of the labeled image $f_1(x\,y)$ can be defined as

$$Hi = \sum_{x,y} I\{f1(x,y) = i\} \quad i = 0, \cdots, n-1 \tag{12}$$

where n is the number of different labels produced by the LBP operator and

$$I\{A\} = \begin{cases} 1 \; A \; is \; true \\ 0 \; A \; is \; false \end{cases} \tag{13}$$

This histogram contains information about the distribution of the local micropatterns, such as edges, spots, and flat areas, over the whole image. For efficient face representation, retaining the spatial information is required; hence, the image is divided into regions R_0, R_1, ..., R_{m-1}, as shown in Figure 7.

Figure 7. An example of a facial image divided into 8 × 8 windows.

The spatially enhanced histogram is defined as

$$H_{i,j} = \sum_{x,y} I\{f_1(x,y) = i\} I\{(x,y) \in R_j\} \quad , \quad \begin{matrix} i = 0, \cdots, n-1 \\ j = 0, \cdots, m-1 \end{matrix} \tag{14}$$

In this histogram, a description of the face on three different levels of locality exists: the labels for the histogram contain information about the patterns on a pixel level, the labels are summed over a small region to produce information on a regional level, and the regional histograms are concatenated to build a global description of the face.

Although Ahnon et al. [48] have mentioned several dissimilarity measures such as histogram intersections, log-likelihood statistics, and Chi square statistics, they used nearest neighbor classifier in their study.

When the image has been divided into several regions, it can be expected that some of the regions contain more useful information than others in terms of distinguishing between people, such as eyes [49, 50]. In order to contribute such information, a weight can be set for each region based on the level of information it contains.

2.5 The proposed LBP-based face recognition

In this article, the proposed face recognition system uses LBP in order to obtain different PDF of each face in different color channels. Each face images after being equalized and segmented will be divided into sub-images in different color channels. For each sub-image in a specific color channel, the PDF will be calculated. The concatenation of these PDFs will result into a single PDF. Due to high correlation of information of *I* color channel with *YCbCr* color channels, only PDFs of *HSI* color space is used for recognition process by using KLD as it was mentioned earlier. Because several decisions have been obtained from different color channels, the combination of these decisions will boost the final decision. In [51], various decision fusion, data fusion, and source fusion techniques have been studied. In this study, sum rule, median rule, majority voting, and feature vector fusion (obtained by concatenating the PDFs before starting the recognition process) have been used in order to combine the multi decisions obtained from LBP–PDF-based face recognition in *HSI* color channels. Table 5 shows the correct recognition of the aforementioned fusion techniques for the HP face database with 15 subjects and 10 samples per each subject, FERET face database with 50 subjects and 10 samples per each subject, and Bosphorus face database with 105 subjects and 32 samples per each subject.

In order to show the superiority of the proposed method on available state-of-the-art and conventional face recognition systems, we have compared the recognition performance with conventional PCA-based face recognition system and the state-of-the-art techniques such as, NMF [52, 53], supervised INMF [54], conventional gray-scale LBP [45], and LDA-based face recognition systems [3] for the FERET face database. The experimental results are shown in Table 6.

The results clearly indicate that this superiority is achieved by using PDF-based face recognition in different color channels backed by the data fusion techniques. Compared to the state-of-the-art technique proposed by Demirel and Anbarjafari [8], the recognition rate obtained by using LBP has slightly better performance.

Table 5. Performance of different decision-making techniques for the proposed face recognition system

	# of training image	SUM RULE	MEDIAN RULE	MV	FVF
H	1	82.82	83.37	80.36	79.94
P	2	95.68	95.55	91.75	86.09
	3	96.45	96.31	92.93	95.25
	4	96.84	97.14	95.02	96.85
	5	98.08	98.38	98.70	97.32
F	1	76.63	75.90	75.09	83.32
E	2	86.49	87.74	87.79	89.43
R	3	90.76	89.41	87.72	97.13
E	4	95.81	93.56	92.00	99.01
T	5	96.36	97.17	94.34	99.78
B	1	74.85	74.94	77.54	79.04
O	7	88.60	88.44	88.63	89.81
S	13	89.97	89.32	89.77	90.10
	19	90.07	92.04	93.01	93.12
	25	91.91	94.07	94.31	95.67

The median rule and FVF-based results are 97.17 and 99.78% for the FERET face database, when five samples per subject are available in the training set, respectively. These results are significant, when compared with the recognition rates achieved by conventional PCA and LDA and the state-of-the-art techniques such as LBP, NMF, and INMF-based face recognition system.

Table 6. Performance of the proposed LBP based face recognition system using FVF, PCA, LDA, conventional gray scale LBP, PDF based face recognition, NMF, and INMF based face recognition system for the FERET face databases with 50 subjects and 10 samples per each subject

# of training images	1	2	3	4	5
FVF	83.32	89.43	97.13	99.01	99.78
FVF [8]	82.89	87.00	96.57	98.80	99.33
PCA	44.00	52.00	58.29	66.17	68.80
LDA	61.98	70.33	77.78	81.43	85.00
LBP	50.89	56.25	74.57	77.67	79.60
NMF	61.33	64.67	69.89	77.35	80.37
INMF	63.65	67.87	75.83	80.07	83.20

Figure 8 shows the FAR and FRR analysis of the proposed face recognition system for the FERET face database. The equal error rate occurs when the recognition rate is about 3.2% which shows that the system can be used in a practical scenario.

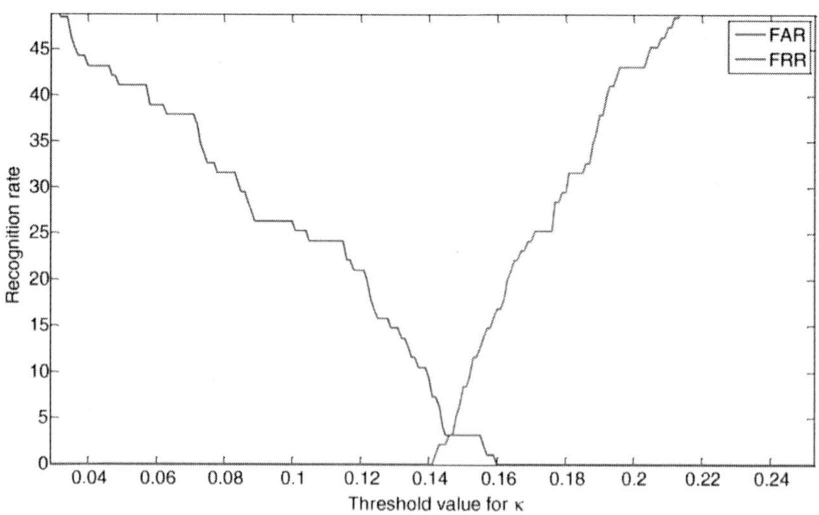

Figure 8. FAR and FRR curves.

3. CONCLUSION

In this article, we have studied a high performance frontal homogenous illumination robust face recognition system using LBP and PDFs in different mutually independent color channels. A face localization which is a combination of local SMQT technique followed by skin tone-based face detection method was employed in this study. DWT + SVD-based image illumination enhancement technique was also applied in order to reduce the effect of illumination. The article analytically analyzed and justified the use of KLD and PDF with a bin size of 256 which was introduced and used in [6, 8]. Several well-known decision fusion techniques have been used in order to combine the decisions obtained from mutually independent color channels. Also an FAR and FRR analysis has been done in this study. Finally, comparison between the proposed method and the conventional and the state-of-the-art techniques has been done which showed the superiority of the proposed method.

ACKNOWLEDGMENTS

The author would like to thank Asst. Prof. Dr. Mikael Nilsson from Blekinge Institute of Technology, for providing the algorithm for local SMQT-based face recognition. Also the author would like to thank Prof. Dr. Ivan Selesnick from Polytechnic University for providing the DWT codes in MATLAB.

COMPETING INTERESTS

The authors declare that they have no competing interests.

REFERENCES

1. Bledsoe WW: *The Model Method in Facial Recognition, Technical report pri 15*. Panoramic Research, Inc, Palo Alto, California; 1964.

2. Turk M, Pentland A: Face recognition using eigenfaces. In *Proceeding of International Conference On Computer Vision and Pattern Recognition*. Maui, Hawaii: IEEE; 1991:586-591.

3. Belhumeur P, Hespanha J, Kreigman D: Eigenfaces vs. Fisherfaces: Recognition using Class Specific Linear Projection. *IEEE Trans Pattern Anal Mach Intell* 1997, 19: 711-720.

4. Liu C, Wechsler H: Comparative assessment of independent component analysis (ICA) for face recognition. In*Proceedings of the Second International Conference on Audio- and Video-based Biometric Person Authentication*. Washington D.C., USA; 1999:211-216. 22-24 March

5. Bartlett MS, Movellan JR, Sejnowski TJ: Face Recognition by Independent Component Analysis. *IEEE Trans Neural Netw* 2002, 13(6):1450-1464.

6. Demirel H, Anbarjafari G: Pose Invariant Face Recognition using Probability Distribution Functions in Different Color Channels. *IEEE Signal Process Lett* 2008, 15: 537-540.

7. Marcel S, Bengio S: Improving face verification using skin color information. In *Proceeding 16th International Conference on Pattern Recognition*. Society Press, Quebec, Canada; 2002:378-381.

8. Demirel H, Anbarjafari G: Data fusion boosted face recognition based on probability distribution functions in different colour channels. *EURASIP J Adv Signal Process* 2009, 2009: 1-10.

9. Anbarjafari G, Demirel H: *Face Recognition Using Colour Statistics*. VDM Verlag: Biometric Recognition; 2010.

10. Laptev I British Machine Vision Conference. In *Improvements of object detection using boosted histograms*. Edinburgh, UK; 2006:949-958.

11. Tae-WoongYoo I-SO: A fast algorithm for tracking human faces based on chromatic PDFs. *Pattern Recognit Lett*1999, 20: 967-978.

12. Rodriguez Y, Marcel S Proceedings of the 9th European Conference on Computer Vision. In *Face authentication using adapted local binary pattern histograms*. Graz, Austria; 2006:321-332.

13. Yang MH, Kriegman D, Ahuja N: Detecting faces in images: A survey. *IEEE Trans Pattern Anal Mach Intell* 2002, 24: 34-58.

14. Chai D, Ngan KN: Face segmentation using skin color map in videophone application. *IEEE Trans Circ Syst Video Tech* 9th edition. 1999, 551-564.

15. Viola P, Jones M: Robust real-time face detection. *Int J Comput Vis* 2004, 52(2):137-154.

16. Nilsson M, Nordberg J, Claesson I IEEE International Conference on Acoustics, Speech, and Signal Processing. In *Face detection using local SMQT features and split up snow classifier*. 2nd edition. Hawaii, USA; 2007:589-592.

17. *A face detector software written by Mike Nilsson, provided in MathWorks exchange file*. 2012. http://www.mathworks.com/matlabcentral/fileexchange/loadFile.do?objectId=13701

18. Gourier N, Hall D, Crowley JL: Estimating Face Orientation from Robust Detection of Salient Facial Features. In*Proceedings of Pointing 2004, ICPR, International Workshop on Visual Observation of Deictic Gestures*. Cambridge, UK;

19. Philips PJ, Moon H, Rizvi SA, Rauss PJ: The FERET Evaluation Methodology for Face Recognition Algorithm.*IEEE Trans Pattern Anal Mach Intell* 2000, 22(10):1090-1104.

20. Savran A, Alyüz N, Dibeklioğlu H, Celiktutan O, Gökberk B, Sankur B, Akarun L: Bosphorus database for 3D face analysis. In *The First COST 2101 Workshop on Biometrics and Identity Management (BIOID 2008)*. Denmark: Roskilde University; 2008:47-56.

21. Kim TK, Paik JK, Kang BS: Contrast enhancement system using spatially adaptive histogram equalization with temporal filtering. *IEEE Trans Consum Electron* 1998, 44(1):82-86.

22. Weeks AR, Sartor LJ, Myler HR: Histogram specification of 24-bit colour images in the colour difference (C-Y) colour space. *Proc SPIE* 1999, 3646: 319-329.

23. Sun CC, Ruan SJ, Shie MC, Pai TW: Dynamic contrast enhancement based on histogram specification. *IEEE Trans Consum Electron* 2005, 51(4):1300-1305.

24. Demirel H, Anbarjafari G, Jahromi MNS. Image equalization based on singular value decomposition. In *23rd IEEE International Symposium on Computer and Information Sciences*. Istanbul, Turkey: Istanbul, Turkey; 2008:1-5. 27-29 October

25. Tian Y, Tan T, Wang Y, Fang Y. Do singular values contain adequate information for face recognition? *Pattern Recogn* 2003, 36: 649-655.

26. Ozcinar C, Demirel H, Anbarjafari G: Satellite Image Contrast Enhancement Using Discrete Wavelet Transform and Singular Value Decomposition. *IEEE Geosci Rem Sens Lett* 2010, 7(2):334-338.

27. Kahraman F, Kurt B, Gokmen M: Robust face alignment for illumination and pose invariant face recognition. In*IEEE Conference on Computer Vision and, Pattern Recognition*. Minnesota, USA; 2007:1-7. 17-22 June

28. Abdullah-Al-Wadud M, Kabir MH, Dewan A, Chae O: A Dynamic Histogram Equalization for Image Contrast Enhancement. *IEEE Trans Consum Electron* 2007, 53(2):593-600.

29. Sung KK, Poggio T: Example based learning for view based human face detection. *IEEE Trans Pattern Recogn Mach Intell* 1998, 20: 39-51.

30. Matsumoto Y, Zelinsky A: Real-time stereo face tracking system for visual interface. In *International Workshop on Recognition, Analysis, and Tracking of Faces and Gestures in Real-Time*. Corfu, Greece; 1999:77-82.

31. Schwerdt K, Crowley J: Robust face tracking using colour. In *4th IEEE International Conference on Automatic Face and Gesture Recognition*. Grenoble, France; 2000:90-95.

32. Huicheng Z, Hongmei L, Daoudi M IEEE Conference on Multimedia and Expo. In *Blocking objectionable images: adult images and harmful symbols*. Lausanne, Switzerland; 2004:1223-1226.

33. Singh SK, Chauhan DD, Vatsa M, Singh R: "A robust skin colour based face detection algorithm". *Tamkang J Sci Eng* 2003, 6(4):227-234.

34. Kjeldsen R, Kender J: Finding skin in colour images. In *Proceedings of the Second International Conference on Automatic Face and Gesture Recognition*. Vermont, USA; 1996:312-317. 14-16 October

35. Anbarjafari G: *A new face recognition system based on colour statistics*. MS Thesis: Eastern Mediterranean University (EMU), Gazimağusa, TRNC; 2008.

36. Osuna E, Freund R, Girosi F: Training support vector machines: an application to face detection. In *Proceedings of the IEEE Conference on Computer Vision and Pattern Recognition (CVPR '97)*. San Juan, Puerto Rico; 1997:130-136.

37. Romdhani S, Torr P, Scholkopf B, Blake A 8th IEEE International Conferences on Computer Vision. In*Computationally efficient face detection*. British Columbia, Canada; 2001:695-700.

38. Cortes C, Vapnik V: Support vector networks. *Mach Learn* 1995, 20(3):273-297.

39. Nilsson M, Dahl M, Claesson I IEEE International Conference on Acoustics, Speech, and Signal Processing (ICASSP). In *The successive mean quantization transform*. Philadelphia, USA; 2005:429-432.

40. Roth D, Yang M, Ahuja N Advances in Neural Information Processing Systems. In *A snow based face detector*. Edited by: Solla SA, Leen TK, Muller K-R. MIT Press; 2000:855-861.

41. Ojala T, Pietikainen M, Harwood D: A comparative study of texture measures with classification based on featured distributions. *Pattern Recogn* 1996, 29: 51-59.

42. Ojala T, Pietikainen M, Maenpaa T: Multiresolutiongray-scale and rotation invariant texture classification with local binary patterns. *IEEE Trans Pattern Anal Mach Intell* 2002, 24: 971-987.

43. Heikkila M, Pietikainen M, Heikkila J British Machine Vision Conference. In *A texture based method for detecting moving object*. London, England; 2004:187-196.

44. Turtinen M, Pietikainen M, Silven O: Visual characterization of paper using isomap and local binary patterns.*Conference Mach Vision Applic* 2005, E-89-D: 1-8.

45. Takala V, Ahonen T, Pietikainen M: Block-based methods for image retrieval using local binary patterns. In *14th Scandinavian Conference on Image Analysis*. Joensuu, Finland; 2005:882-891.

46. Jin H, Liu Q, Lu H, Tong X: Face detection using improved LBP under Bayesian framework. In *3rd International Conference on Image and Graphics*. Porto, Portugal; 2004:306-309. 18-20 December

47. Zhang W, Shan S, Gao W, Chen X, Zhang H 10th IEEE International Conference on Computer Vision. In *Local Gabor binary pattern histogram sequence: a novel non-statistical model for face representation and recognition*. Beijing, China; 2005:786-791. 17-21 October

48. Ahnon T, Hadid A: Pietikainen, Face recognition with local binary patterns. In *European Conference on Computer Vision*. Prague, Czech Republic; 2004:469-481.

49. Gong S, McKenna SJ, Psarrou A: *Dynamic Vision, From Images to Face Recognition*. London: Imperial College Press; 2000.

50. Zhao W, Chellapa R, Rosenfeld A, Philips PJ: *Face Recognition: A Literature Survey, Technical report CAR-TR-948*. University of Maryland: Centre for Automation Research; 2002.

51. Polikar R: Ensemble Based Systems in Decision Making. *IEEE Circ Syst Mag* 2006, 6(3):21-45.

52. Lee DD, Seung HS: Learning the parts of objects by nonnegative matrix factorization. *Nature* 1999, 401: 788-791.

53. Lee DD, Seung HS: Algorithms for nonnegative matrix factorization. *Adv Neural Inform Process Syst* 2001, 13: 556-562.

54. Chen WS, Pan BB, Fang B, Li M, Tang JL: Incremental nonnegative matrix factorization for face recognition.*Mathematical Problems in Engineering* 2008, 1-17. Vol

Chapter 3

2DPCA fractal features and genetic algorithm for efficient face representation and recognition

Yousra Ben Jemaa*, **Ahmed Derbel and Ahmed Ben Jmaa**

Signals and Systems Unit, National Engineering School of Sfax, Sfax University, BP W 3038, Sfax, Tunisia

ABSTRACT

In this article, we present an automatic face recognition system. We show that fractal features obtained from Iterated Function System allow a successful face recognition and outperform the classical approaches. We propose a new fractal feature extraction algorithm based on genetic algorithms to speed up the feature extraction step. In order to capture the more important information that is contained in a face with a few fractal features, we use a bi-dimensional principal component analysis. We have shown with experimental results using two databases as to how the optimal recognition ratio and the recognition time make our system an effective tool for automatic face recognition.

KEYWORDS

face recognition, fractal coding, 2DPCA, IFS, genetic algorithms

I. INTRODUCTION

The human face is a very rich source of information that can be used to identify persons. This ability of recognition allows us to distinguish persons despite the facial resemblance between them. Nowadays, many researchers try to benefit from computer applications, which become widely used in face automatic recognition.

After more than 30 years of research, we can classify the different existing face recognition systems into three main approaches.

- Local approaches which are based on the fact that the face contains parts that have a high discriminating power such as eyes, nose, mouth... To recognize a person, we use either the blocks containing these regions or the geometric relationships between them [1, 2]. Representative works include hidden Markov model [3], elastic bunch graph matching algorithm [4]...

- There are global approaches which treat the face as a whole object and use all the information included in it. Many methods have been proposed that include the use of Eigenfaces [5], discrete cosine transform, and Gabor Wavelets [6]... These methods suffer from the size of the feature vector provided to the classifier. For this reason, many linear and nonlinear methods for vector size reduction are applied (PCA, LDA, ICA, ...).

- Hybrid approaches: The principle of these approaches is to imitate the human visual system, which uses both local and global features to recognize persons. The combination of these two methods has only one interest: to take advantage of the combined benefits of both approaches [7, 8].

Despite the number of researchers and the proposed methods, several factors can significantly affect face recognition performances, such as the pose, the presence/absence of structural components, facial expressions, occlusion, and illumination variations.

In order to encounter these factors and ensure a high recognition rate and a fast recognition time, we have used, in this article, the fractal representation which exploits the inter-image resemblance [9]. There are few articles that are related to this topic [face recognition using Iterated Function System (IFS) theory] [10–14]. A description of some of these studies and their differences from the proposed method can be found in Section 6.

The proposed system contains the following steps:

- Normalization of the original image.
- Feature extraction using fractal encoding of the normalized image and genetic algorithm.
- Application of the bi-dimensional principal component analysis (2DPCA) technique on the fractal code to reduce the feature vector dimension.
- Classification using Multi layer perceptron.
- The idea proposed in this article has two major advantages compared with the other approaches:
- Reduced size of the fractal code represents the feature vector. Since it has a reduced dimension, the recognition can be ensured with satisfactory time. We have proposed a new fractal algorithm based on genetic algorithm to ensure a low time for feature extraction step.
- High fidelity compared with the original image. The fractal code represents discriminant features of the original image. These features are invariant overlooked lighting, rotation, and translation of the face and scaling, because the IFS theory takes into account these variations.

We proposed to apply a 2DPCA to represent face by a few fractal features having a high discriminatory power.

This article is organized as follows: Basic notions concerning IFS, fractal coding theory and the new fractal algorithm based on genetic algorithm are provided in Section 2. Fractal features are presented in section 3. The most discriminating fractal parameters extracted using 2DPCA are described in Section 4. Section 5 provides face recognition system based on neural networks, the experimental results and Comparison between the two types of features obtained using IFS and PCA-IFS, respectively. A comparison with other approaches is also done in section 6. Conclusion and future works are presented in Section 7.

2. GENETIC ALGORITHM FOR FRACTAL CODING

2.1. IFS theory

The IFS theory is proposed by Barnsley, who suggested that, instead of storing all the pixels of the still image, we can keep only a collection of global contracting transformations such as rotation and contrast scaling [15].

Image fractal encoding is well known in the literature. It has been widely used for image compression [9, 16]. In this article, we have used it for classification purpose.

Firstly, the coding involves the partitioning of the image into ranges R_i, which do not intersect and can have fixed size or not (quadtree partitioning), and domain D_i which can intersect. Secondly, we have searched the best range/domain matching by applying a transformation W_i to each domain D_i (see Figure 1).

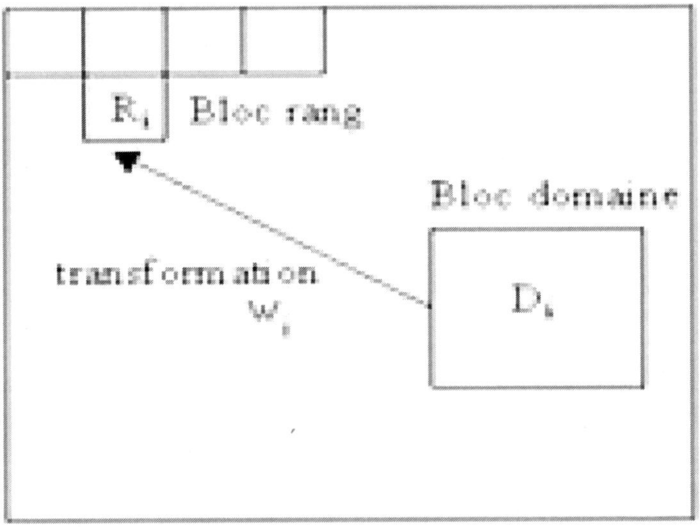

Figure 1. Range/Domain matching.

This is possible because fractal coding is based on the self-similarity of the face, which means that regions can be the transformed versions of some others like shown in Figure 2.

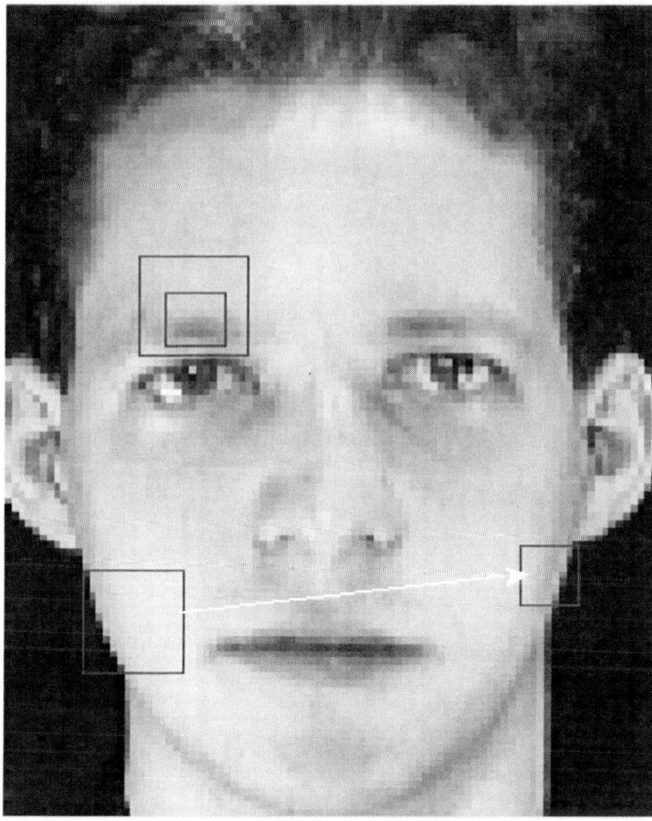

Figure 2. Inter image similarity.

Therefore, to code an image, we need to determine a set of R_i, D_i, and W_i. To achieve an excellent coding phase, we should make a good choice of transformation W_i between both R_i and D_i. Then, we have to find the perfect adjustment of the contrast S_i and the lighting O_i for each W_i using the method of least square [9].

2.2. The proposed algorithm

The major problem of standard fractal coding is time consumption compared with other methods of image coding. The time is essentially spent on the search of the similar domain block. We present in this article, a new genetic algorithm for image coding, that speeds up this method. In the next, we have detailed our algorithm: the representation of the fitness function, the Genetic operators and some other improvements to the simple genetic algorithms.

There are many algorithms of optimization used for different domains. We have chosen genetic algorithm [17–19] to accelerate our fractal image coding algorithm. We have given details of genetic characteristics in the following section.

1) Chromosome attributes

According to the regions parameter coding, a chromosome is constituted by *N* genes, where *N* is the number of regions not yet coded.

The gene is composed of three parameters (X_{Dom}, Y_{Dom}), that represent the domain block coordinates and the rotation W_i. These three parameters are integers.

- $X_{Dom} \in [0, L]$, *L* is the image length.
- $Y_{Dom} \in [0, W]$, *W* is the image width.
- $W_i \in [0, 7]$, eight possible rotations.

Figure 3 illustrates a chromosome representation.

Region 1	X_{Dom}^{1}	Y_{Dom}^{1}	$Flip^{1}$
Region 2	X_{Dom}^{2}	Y_{Dom}^{2}	$Flip^{2}$
	⋮	⋮	⋮
Region N	X_{Dom}^{N}	Y_{Dom}^{N}	$Flip^{N}$

Figure 3. Chromosome representation.

2) Genetic operators

The crossover and mutation operators ensure the production of offspring. These genetic operators must be defined according to the chromosome specification. With these basic components, a genetic algorithm works as follows: The first procedure is to generate the first population represented with string codification (chromosome) that represents possible solution to the problem. Each individual is evaluated, and according to its fitness, an associated probability to be selected for reproduction is assigned.

- The crossover operator combines two individuals (the parents) of the current generation whose chromosomes have not given selected solution to produce two offspring individuals. According to our chromosome specification, a new scheme of the crossover operator is proposed. The offspring coordinates and the isometric flip are selected randomly from the parents as presented in Figure 4.

- Mutation operator modifies the chromosome genes randomly according to the mutation probability. Genes (X_{Dom} , X_{Dom}, W_i) are changed with random generated values, respectively, in [0, L], [0, W], and [0, 7] intervals (see Figure 5).

3) Fitness measure

The fitness function assigns to each individual in the population a numeric value, that determines its quality as a potential solution. The fitness denotes the individual's ability to survive and to produce offspring.

In our case, the fitness is the number of regions that can be coded with root mean square error (RMSE)less than a fixed value. The RMSE is the distance between the region and the domain block is determined by its coordinates (X_{Dom} , X_{Dom}) and transformed with corresponding contrast *S* and the lighting *O*.

The RMSE parameter is given in the following:

$$\text{RMSE} = \|S.D_i - (R_j - O)\| \tag{1}$$

where || . || is the two norm function, D_i is domain elements, R_i denotes the range elements, and values of contrast*S* and lighting *O* are obtained when minimizing the RMSE criterion (they are the two arguments that minimize the RMSE).

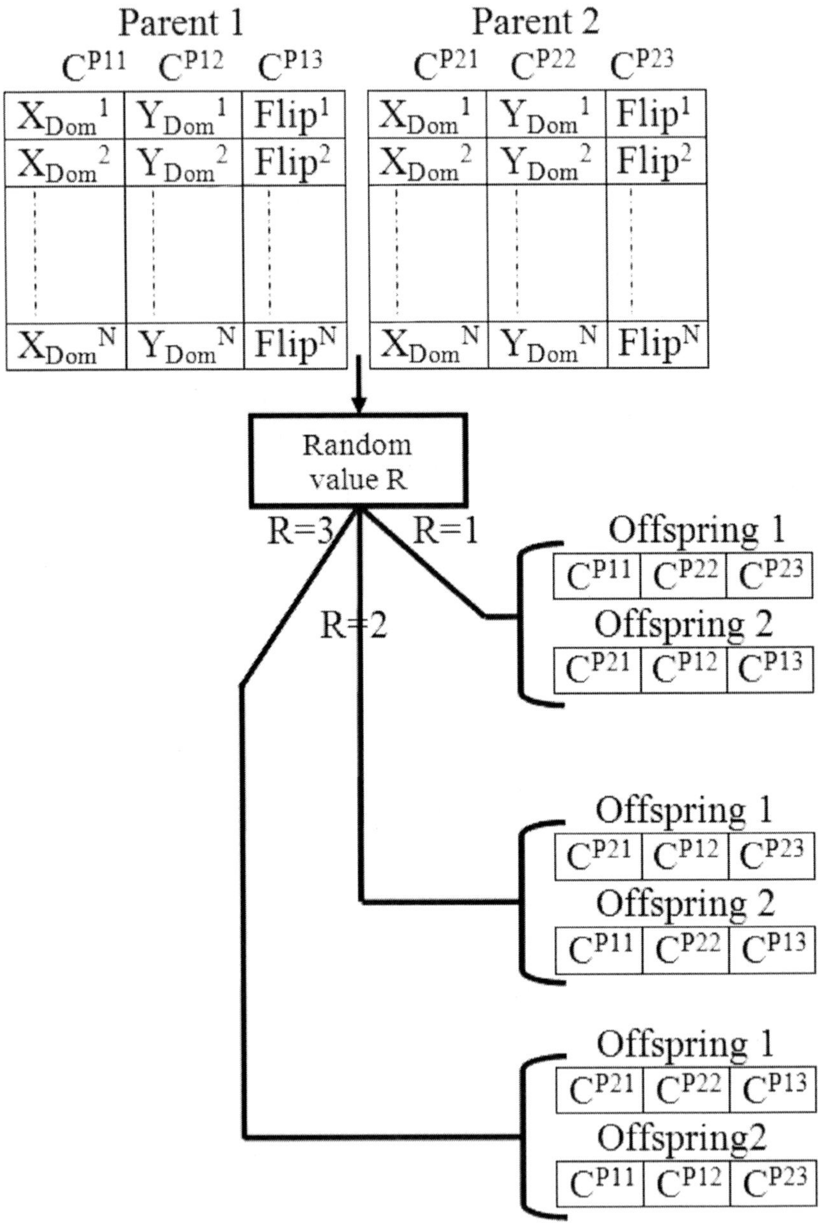

Figure 4. Crossover operator scheme.

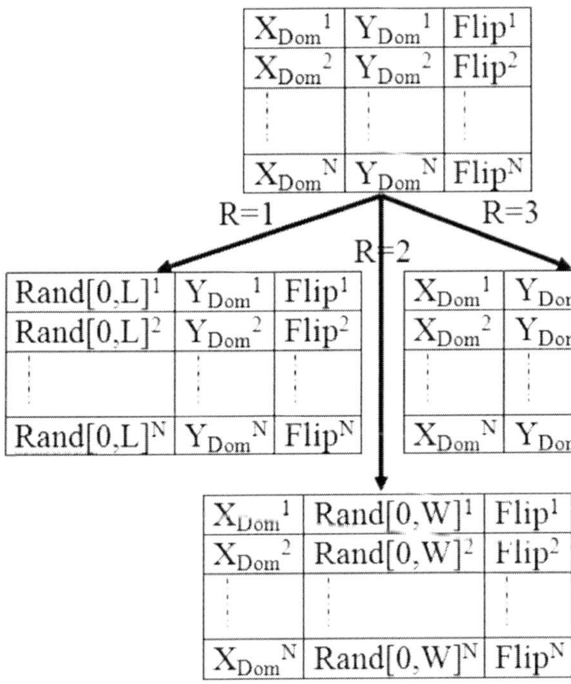

Figure 5. Mutation operator scheme.

4) Genetic coding algorithm

Genetic algorithms have been used previously to find solutions to the minimization problems related to the fractal inverse problem [18]. Here, we describe the Genetic Algorithm that we have used to speed up the coding algorithm. This algorithm is used for all decomposition schemes. In spite of the range block size and position, the domain block is always double the size of the range one.

(Input I: NxN gray scale image [Image would be square] Output W: Coded IFS);

(Region Size) = 16; (Fixed Error) = X;

Decompose the input image Into (Region Size) blocks;

While Exist (Regions not coded)

Scale the Domain Blocks;

Generate a random population of chromosomes;

While Exist (Regions not coded) and (Last generation not reached)

- Compute fitness for all regions;
- When optimal domain block found write obtained transformation parameters to the output W;
- Generate new population Apply Crossover and Mutation operators; Wend

(RegionSize) = (RegionSize)/2;

If Regions size > 4

- Decompose the rest region not coded into (Range Size) blocks;

Else

- (FixedError) = (FixedError) + X;
- Code all remaining Regions;

IEnd

Wend

3. FRACTAL FEATURES EXTRACTION

After fractal coding, where each domain is compared with all regions of the image, we obtain a set of transformations which can approximate the face image. Each transformation is represented by parameters of contrast S_i, brightness O_i, spatial coordinates of Range/Domain, and rotation W_i(seven parameters). The size of the obtained feature matrix is equal to 7× the number of transformations necessary to code all regions. So reducing the size of the information is necessary for minimizing the recognition time. An immediate reduction of the feature vector consists of replacing the coordinates of the regions and domains by two normalized distances:

- *x*: the distance between the Domain D_i and the region R_i according to the abscissas,
- *y*: the distance between the Domain D_i and the region R_i as the ordinates.

The size of the new matrix is then equal to 5× the number of transformations.

Despite all the reductions of the fractal vector, it remains quite large. Thus, we proposed to use a two-dimensional PCA to extract the most discriminating features.

4. THE DISCRIMINATING PARAMETERS OF FRACTAL FEATURES

The 2DPCA is a method of data analysis, based on finding a new reference on which we represent the information while keeping only discriminating data [20]. As opposed to conventional PCA, 2DPCA is based on matrices rather than vectors. Consequently, the covariance matrix can be constructed directly using original matrix of features. So, when using 2DPCA, it is easier to evaluate the covariance matrix, and less time is required to determine the corresponding eigenvectors.

The idea consists of projecting each feature matrix X ($n \times m$) through a linear transformation.

The first step is to calculate the covariance matrix G_t of fractal features which is obtained from the images of the training database as follows:

$$G_t = \frac{1}{M} \sum (X_j - \hat{X})^T * (X_j - \hat{X}), \tag{2}$$

where M is the number of images in the database, X_j represents the fractal matrix obtained from the image number $j$of the training database, and X^ $\hat{X}$ is the average of all fractal matrices associated to the images from the training database.

The next step is to choose d eigenvalues associated with eigenvectors obtained from the previously calculated covariance matrix These eigenvalues determine the new reference that minimizes the criterion $J(R)$ defined by:

$$J(R) = R^T * G_t * R$$
$$R_i^T R_j = 0, i \neq j, i, j = 1, \ldots, d \tag{3}$$

The third step is to extract the main features of X as follows:

$$Y = X * R \tag{4}$$

where $R = [R_1R_2 \ldots R_d]$ is the projection matrix and $Y = [Y_1Y_2 \ldots Y_d]$ is the fractal feature matrix produced after applying 2DPCA.

To project the matrix in the new base, we have selected the eigenvectors associated with the largest eigenvalues. The biggest shortcoming of 2DPCA is the choice of the number of retained eigenvalues. To solve this problem, researchers have adopted different solutions, either heuristically [21] or graphically according to the shape of eigenvalues [22]. In this article, we used a graphically method to select the most important eigenvectors.

5. EXPERIMENTAL RESULTS

5.1. Overview of the used face databases

To highlight the performances of the proposed system, we have carried out the first experiment on the Yale database [23], with the aim of pinpointing the behavior of our approach under changing face expressions and poses. This base contains 165 images of 15 individuals. In this experiment, 30% of all image samples per class are chosen randomly and are used for training, and the remaining images for test. The proposed approach has also been applied on the ORL database [24], which contains 10 different images of each of the 40 distinct individuals. For this database also, 30% image samples per class are chosen randomly, and are used for training and the remaining images for test. In the ORL database, images are taken at different lighting conditions, facial expressions, and orientations which allows testing the behavior of our approach under these changes.

5.2. The classification system

The face recognition was ensured by a multilayer perceptron architecture. The training of weights is assured by the algorithm of retro-propagation. This architecture is the most used one because it can reduce miss-classification among the neighborhood classes.

5.3. Face recognition using fractal features

In order to have fractal feature vectors with the same length, the size of the face must be normalized (32×32). The normalized image is coded by 64 transformations using fractal code. Consequently, we obtained 320 fractal features as each transformation is coded on 5 parameters, as already explained in Section 3.

Table 1 shows the performance of our system using fractal features for the two databases. Each of the recognition rate has been tested on five random combination of the face samples. According to these observations, fractal features gave very good results.

Table 1. Recognition rate using fractal features

Database	Recognition rate
Yale	99.16
ORL	98.33

5.4. Face recognition using 2DPCA-IFS features

Given the large number of the used parameters (320 parameters), our idea was to apply a bi-dimensional PCA on the fractal features for the two already-mentioned databases. After a 2DPCA was applied on all features matrices, we were able to achieve the following results found in Table 2.

Table 2. Recognition rate versus the number of transformations

Number of transformations	RR(Yale)	RR(ORL)
2	66.67	42 .33
5	98.15	97.15
8	98.15	97.29
12	98.67	97.29
15	98.67	97.51
30	98.67	97.66
45	98.83	97.66
60	98.83	97.67

We study here the relation between the number of transformation used as features and the recognition rate. We have found that recognition rate grows until transformation number reaches a value about 5, and then it remains almost constant as transformation number continues to grow. Consequently, the compromise recognition rate/transformation (the optimal situation is given with the minimum number of transformation and a satisfactory recognition rate) number is solved for five eigenvectors for the two databases. This can be explained according to the variation of the eigenvalues. We can keep until the fifth eigenvalue and the remaining eigenvalues can be neglected as shown in Figure 6. Here, we preserve all the five parameters for each transformation, for the two databases.

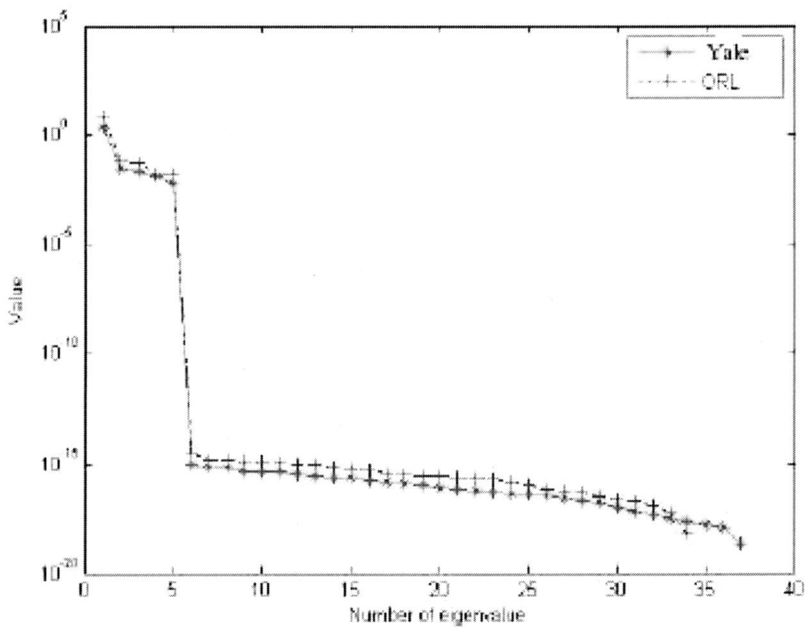

Figure 6. Eigenvalues representation in descending order.

From the previous analysis, we can notice that the best choice to keep is five transformations where each one is coded by five parameters to ensure a good recognition phase.

5.5. Comparison between IFS and 2DPCA-IFS features

In Table 3, we present the recognition rate obtained, when using all fractal features, and those reduced by the 2DPCA with the optimal configuration (five transformations where each one is coded by five parameters). While comparing the two methods, we can deduce those recognition rates that were found using the two databases which did not decrease significantly.

Table 3. Recognition rate for the two approaches and the two databases

Approach	Yale	ORL
IFS and GA	99.165	98.33
2DPCA-IFS and GA	98.15	97.15

The major advantage of 2DPCA-IFS method is that the number of parameters decreases from 320 parameters with IFS to only 25 parameters with 2DPCA-IFS, which can reduce the recognition time while keeping a very satisfactory recognition rate.

5.6. Comparison with other approaches

Here we compare our results with earlier results published in [10–12]. ORL database is used for comparison. The FND method [10] consists of three steps:

- A standard fractal coding giving a code for each image in the database.
- Each image I is decoded with each code in the database to generate the output image called the attractor.
- A classification step for each test image I using the minimization of FND distance d_{FN} (fractal neighbor distance) defined as the distance between the test image and the attractor image.

$$d_{FN} = d(f_j(I), I), \tag{5}$$

where f_j is the j th fractal code in the database, $f_j(I)$ is the decoded image using the code f_j.

The LR-SNN-T method [11] is based on an equalization of the original image and a normalization of its dimension using a bicubic interpolation. The feature vector is represented by the whole image after processing. The classification is achieved by a multilayer perceptron.

Finally, the X method [12] consists on extracting parts of face, containing the most discriminating information like eyes, nose... Then applying a standard fractal coding on each detected part, the classification is also ensured by a multilayer perceptron.

Although both approaches, the FND and X, use the IFS coding, they are completely different from our approach. This difference is summarized in Table 4.

Table 4. Differences between our approach and other approaches

Approach	FND	X	Our approach
Technique used for feature extraction	Standard fractal decoding	Standard fractal coding applied in facial regions	Genetic algorithm for fractal coding
Classification	FND distance	MLP	MLP
Databases used for evaluation	ORL-Yale	ORL–In room database	ORL–Yale

Table 5. Recognition rates/times for different methods

Approach	IFS	IFS and GA	2DPCA-IFS and GA	FND	LR-SNN-T	X
Recognition rate	98.4	98.33	97.15	95.25	90.25	85
Recognition time (S)	7.59	1.75	1.41	9.1	38.12	n/a

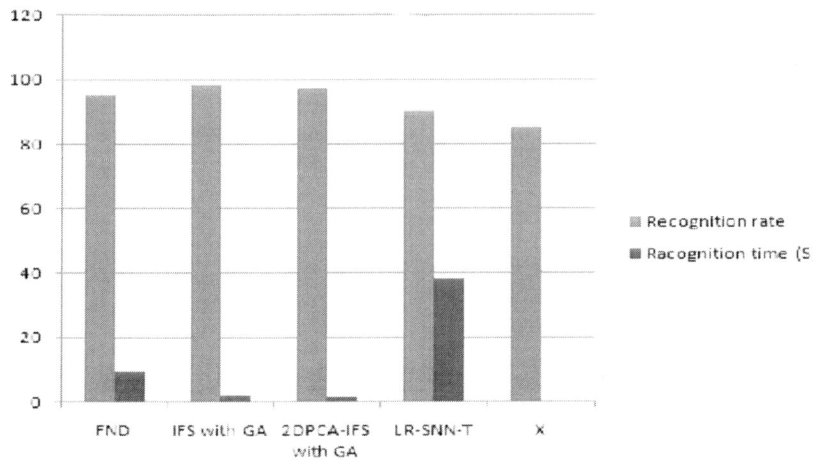

Figure 7. Recognition rates/times for different methods.

The performance comparison between all approaches is shown in Table 5 and Figure 7.

We conclude that

- Fractal features are much more powerful than others and are good means to characterize faces.
- Five eigenvalues are sufficient to code faces. A little improvement is observed when more than five eigenvalues are used.
- The robustness of our 2DPCA-IFS approach is that it gives the best time recognition, and thanks to the use of genetic algorithm and 2DPCA technique, which keeps a high recognition rate proving its applicability for real time system.

6. CONCLUSION

A hybrid approach is introduced in which, through the 2DPCA, the most discriminating genetic fractal features are extracted and used as the input of a neural network.

The performance of our method is both due to the fidelity of fractal coding for representing images, the genetic algorithm to speed up the features extraction step, and the 2DPCA which highlights all discriminating features.

Compared with other approaches, the proposed recognition method has achieved high recognition rate and low recognition time for the two databases.

COMPETING INTERESTS

The authors declare that they have no competing interests.

REFERENCES

1. Ben Jemaa Y, Khanfir S: **Automatic local Gabor features extraction for face recognition.** *Int J Comput Sci Inf Secur* 2009, **3**(1):116-122.

2. Cox IJ, Ghosn J, Yianilos PN: **Feature-based recognition using mixture-distance.** *IEEE Conference on Computer Vision and pattern Recognition* 1996, 209-216.

3. Nefian AV, Hayes MH: **An embedded HMM based approach for face detection and recognition.** *IEEE International Conference on Acoustics, Speech and Signal Processing* 1999, **6:** 3553-3556.

4. Wiskott L, Fellous JM, Kruger N: **CVD Malsburg, Face recognition by elastic bunch graph matching.***Intelligent Biometric Techniques in Fingerprint and Face Recognition* 1999, **Chapter 11:** 355-396.

5. Moon H, Philips PJ: **Computational and performance aspects of PCA based face recognition algorithms.***Perception* 2001, **30:** 303-321.

6. Hang HZ, Zhang B, Huang W, Tian Q: **Gabor wavelet associative memory for face recognition.** *IEEE Trans Neural Netw* 2005, **16**(1):275-278.

7. Zhao W, chellappa R, Phillips PJ, Rosenfeld A: **Face recongnition: a literature survey.** *ACM Comput Surv*2003, **35**(4):399-458.

8. Chen CJ: **Integration of local and global features for face recognition.** *International Conference on Neural Networks and Signal Processing* 2008, 193-198.

9. Wohlberg BE, De Jager G: **A review of the fractal image coding literature.** *IEEE Trans Image Process* 1999,**8**(12):1716-1729.

10. Tan T, Yan H: **Object recognition based on fractal neighbor distance.** *Signal Process* 2001, **81:** 2105-2129.

11. Zeb J, Javed MY, Qayyum U: **Low resolution single neural network based face recognition, in Proceedings of World Academy of Science.** *Engineering and Technology* 2007., **22:**

12. Temdee P, Khawparisuth D, Chamnongthai K: **Face recognition by using fractal encoding and backpropagation neural network.** *International Symposium on Signal Processing and its Applications, Brisbane, Australia* 1999.

13. Abate AF, Nappi M, Riccio D, Tucci M: **Occluded face recognition by means of the IFS.** *Lecture Notes in Computer Science* 2005., **3656:**

14. AF, Nappi M, Riccio D, Tortora G: **An IFS based approach for face recognition.** *IEEE ICIP* 2005., **2:**

15. Barnsley B, Hurt L: *Fractal Image Compression.* Peters, Wellesley; 1993.

16. Fisher Y: *Fractal Image Compression with Quadtrees.* book Springer-Verlag London, UK; 1995.

17. Goldberg DE: *Genetic Algorithms in Search, Optimization, and Machine Learning.* Addison Wesley Publishing Company; 1989.

18. Shonkwill R, Mendivil F, Deliu A: **Genetic algorithms for the 1-D fractal inverse problem.** *Proceedings of the Fourth International Conference on Genetic Algorithms, San Diego* 1991.

19. Wu M-S, Teng W-C, Jeng J-H, Hsieh J-G: **Spatial correlation genetic algorithm for fractal image compression.** *Fractals* 2006, **28**(N 2):497-510.

20. Yang J, Zhang D, Frangi AF, YuYang J: **Two-dimentional PCA: a new approach to appearence-based face representation and recognition.** *IEEE Trans Pattern Anal Mach Intell* 2004, **26:** 131-137.

21. Turk MA, Pentland A: **Face recognition using eigenfaces.** *IEEE conference on Computer Vision and Pattern Recognition* 1991, 586-590.

22. Turk MA, Pentland AP: **Eigenfaces for recognition.** *J Cogn Neurosci* 1991, **3**(1):71-86.

23. Georghiades A, Kriegman D, Belhumeur P: **From few to many: generative models for recognition under variable pose and illumination.** *IEEE Trans Pattern Anal Mach Intell* 2001, **40:** 643-660.

24. **ORL Face Database**[http://www.cl.cam.ac.uk/research/dtg/attarchive/facedatabase.html]

Chapter 4

Automatic Face Recognition System for Hidden Markov Model Techniques

Peter M. Corcoran[1] and Claudia Iancu[1]

[1] College of Engineering & Informatics,National University of Ireland Galway, Ireland

1. INTRODUCTION

Hidden Markov Models (HMMs) are a class of statistical models used to characterize the observable properties of a signal. HMMs consist of two interrelated processes: (i)an underlying, unobservable Markov chain with a finite number of states governed by a state transition probability matrix and an initial state probability distribution, and (ii) a set of observations, defined by the observation density functions associated with each state. In this chapter we begin by describing the generalized architecture of an automatic face recognition (AFR) system. Then the role of each functional block within this architecture is discussed. A detailed description of the methods we used to solve the role of each block is given with particular emphasis on how our HMM functions. A core element of this chapter is the practical realization of our face recognition algorithm, derived from EHMM techniques. Experimental results are provided illustrating optimal data and model configurations. This background information should prove helpful to other researchers who wish to explore the potential of HMM based approaches to 2D face and object recognition.

2. FACE RECOGNITION SYSTEMS

In this section we outline the basic architecture of a face recognition system based on Gonzalez's image analysis system [Gonzalez & Woods 1992] and Costache's face recognition system [Costache 2007]. At a top-level this is represented by the functional blocks shown in Figure 1.

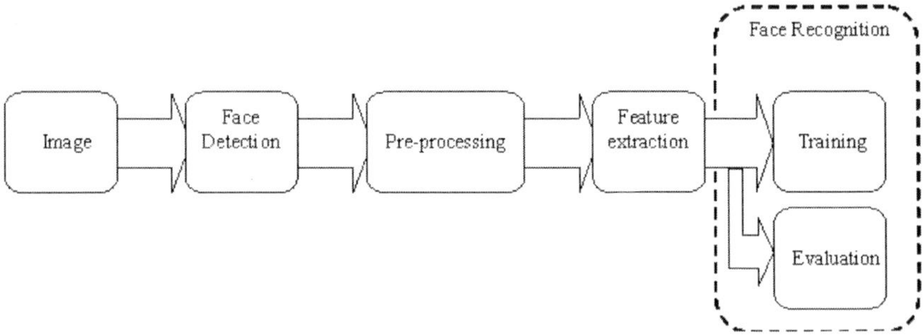

Figure 1. The architecture of a face recognition system

1 **Face detection and cropping block**: this is the first stage of any face recognition system and the key difference between a semi-automatic and a fully automatic face recognizer. In order to make the recognition system fully automatic, the detection and extraction of faces from an image should also be automatic. Face detection also represents a very important step before face recognition, because the accuracy of the recognition process is a direct function of the accuracy of the detection process [Rentzeperis *et. al.* 2006, Corcoran *et. al.* 2006].

2 **Pre-processing block**: the face image can be treated with a series of pre-processing techniques to minimize the effect of factors that can adversely influence the face recognition algorithm. The most critical of these are *facial pose* and *illumination*. A discussion on these factors and their significance w.r.t. HMM techniques is given in Section 3.

3 **Feature extraction block**: in this step the features used in the recognition phase are computed. These features vary depending on the automatic face recognition system used. For example, the first and most simplistic features used in face recognition were the geometrical relations and distances between important points in a face, and the recognition 'algorithm' matched these distances [Chellappa *et. al.* 1992]; the most

widely used features in face recognition are KL or eigenfaces, and the standard recognition 'algorithm' uses either the Euclidian or Mahalanobis distance [Chellappa *et. al.* 1992, 1995] to match features. Our features and the extraction method used are described in Section 4.

4 **Face recognition block**: this consists of 2 separate stages: a *training process*, where the algorithm is fed samples of the subjects to be learned and a distinct model for each subject is determined; and an *evaluation process* where a model of a newly acquired test subject is compared against all exisiting models in the database and the most closely corresponding model is determined. If these are sufficiently close a recognition event is triggered.

3. FACE DETECTION AND CROPPING

As mentioned in the previous section, face detection is one of the most important steps in a face recognition system and differentiates between semi-automatic and fully automatic face recognizer. The goal of an automatic face detector is to search for human faces in a still image and, if found, to accurately return their locations. In order to make the detection fully automatic the system has to work without input from the user. Many attempts to solve the problem of face detection exist in the literature beginning with the basic approach of [Kanade 1977] and culminating with the method of [Viola & Jones 2001]. Comprehensive surveys of face detection techniques can be found in [Yang *et. al.*2002] and [Costache 2007]. In this section we underline the main challenges an automatic face detector has to tackle, and we briefly describe the face detector used in our experiments. Face detection methods were classified by [Yang *et. al.* 2002] into four principle categories: (i) knowledge-based, (ii) feature invariant, (iii) template matching and (iv) appearance-based methods. According to [Gonzalez & Woods 1992], the main disadvantage presented by the majority of these methods is the time required to detect all the faces in an image. State-of-the-art face detection methods provide real-time solutions. The best known of these methods, and the gold standard for face detection was originally proposed by [Viola & Jones 2001]. The original algorithm was, according to its authors, 15 times faster than any previous approach. The algorithm has been well proved in recent years as being one of the fastest and most accurate face detection algorithms reported and is presently the gold standard against which other face detection techniques are benchmarked. For these reasons we adopted it to implement our face detection subsystem.

Our implementation of the Viola & Jones detection algorithm is provided in the Intel digital image processing C++ library [OpenCV 2006]. This can be used both for face detection and subsequent cropping of confirmed facial images. The OpenCV face detector has been pre-trained using a very comprehensive database of face/non-face examples and is widely used in the literature.

4. PRE-PROCESSING TECHNIQUES

Automatic face detection is influenced by a number of key factors [Costache 2007]: facial *orientation*or *pose*: the appearance of the face varies due to relative camera-face pose, between full frontal images and side-profile images; *in-situ occlusions* such as facial hair (e.g. beard, moustache), eye-glasses and make-up; facial *expressions* can significantly influence the appearance of a face image;*overlapping occlusions* where faces are partially occluded by other faces present in the picture or by objects such as hats, or fans; *conditions of image acquisition* where the quality of the picture, camera characteristics and in particular the *illumination conditions* can strongly influence the appearance of a face.

For our system to perform better in the recognition stage, we apply a set of pre-processing techniques: the first step in pre-processing is to bring all images into the same color space and to normalize the size of face regions. This normalization process is critical to improving the final face recognition rate and we will later present some experimental results for our HMM-specific AFR.

4.1. Color To Grayscale Conversion

In most face recognition applications the images are single or multiple views of 2D intensity data [Zhao *et. al.* 2003], and many databases built for face recognition applications are available as grayscale images. From the four databases used in our experiments, 3 contained grayscale images (BioID, Achermann, UMIST) and one contained RGB images (FERET). Practical images will, naturally, be acquired in color as modern image acquisition systems are practically all color and so we need to convert from color to grayscale, or intensity images of the selected face regions. In practice the intensity data may be available from the imaging system – many camera system employ YCC data internally and the Y component can be utilized directly. In other cases we may need to perform an explicit conversion of RGB data. Here a set of red, green and blue integer values characterize an image pixel. The effective luminance, Y of each pixel is calculated with the following formula [Pratt 1991]:

$$Y = 0.3 \times Red + 0.59 \times Green + 0.11 \times Blue \quad (1)$$

4.2. Image Resizing

For a HMM-based face recognition system having a consistently sized face region is particularly important because the HMM requires regional analysis of the face with a scanning window of fixed size. A straightforward approach is to resize all determined face regions to a common size. To facilitate more efficient computation we seek the smallest sized face region possible without impacting the overall system recogntion rate. Some empirical data will be presented later to illustrate how different factors, including the size of normalized face regions, affect recognition rate.

There are many techniques that can be used to enlarge or reduce the size of an image. These methods generally realize a trade-off between speed and the degree to which they reduce the occurrence of visual artifacts in the resulting image. The most commonly used resize method is called bicubic interpolation and has the advantage that the interpolated image is smoother than images obtained using simpler interpolation techniques and has fewer artifacts [Lehmann *et. al.* 1999]. In our work we have used bicubic spline interpolation using bicubic polynomials. More details of how to calculate bicubic spline interpolation functions can be found in [Hummel 1977].

4.3. Illumination Normalization

One of the most important factors that influence the recognition rate of a system is illumination variation. In was shown in [Adini *et al.* 1997, Gokemen *et al.* 2007] that variations in illumination can be more relevant than variations between individual characteristics. Such variations can induce an AFR system to decide that two different individuals with the same illumination characteristics are more similar than two instances of the same individual taken in different lighting conditions. Thus normalizing illumination conditions across detected face regions is crucial to obtaining accurate, reliable and repeatable results from an AFR. One approach suitable for face models which combine both facial geometry and facial texture such as active appearance models (AAM) is described by [Ionita 2008]. However as HMM techniques do not explicitly rely on facial geometry or textures it is not possible to integrate the illumination normalization within the structure of the model itself. Instead we must rely on a discrete illumination normalization process. Fortunately most AFR systems employ a similar prefiltering stage and we can draw on a wide range of techniques from the literature.

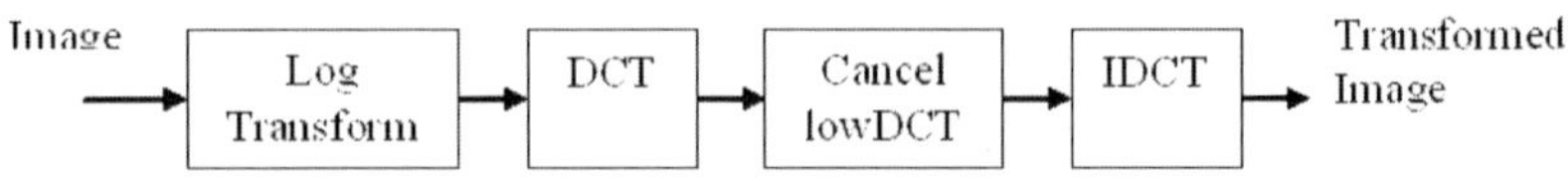

Figure 2. Block scheme of logDCT algorithm

Algorithms used for performing the normalization vary from a simple histogram equalization (HE) to more complex techniques such as albedo maps [Smith & Hancock 2005] and contrast limited adaptive histogram equalization (CLAHE) [Zuiderveld 1994, Pizer *et al* 2006, Corcoran *et al* 2006]. These algorithms perform well when the variations in illumination are small but there is no commonly adopted method for illumination normalization in images which performs well for every type of illumination. Some tests have been conducted to determine the robustness of face recognition algorithms to changes in lighting [Phillips *et al* 2000, O'Toole *et al* 2007]. Also, numerous illumination normalization techniques have been developed. Some of the more widely used of these - *histogram equalization*, *histogram specification* and *logarithm transformation* - have been compared in [Du *et al*2005] with more recently proposed methods, *gamma intensity* correction and *self-quotient image*. The results are interesting: both HE and logarithmic transform improved recognition rates over face regions that were not normalized, compared favorably to the other techniques.

To tackle the problem of illumination variations we implemented the following three illumination normalization algorithms: (i) *histogram equalization* (HE) based on [Gonzalez & Woods 1992], (ii) *contrast limited adaptive histogram equalization* (CLAHE) based on [Zuiderveld 1994], and (iii) the relative new method of DCT in the logarithm domain - *logDCT* based on [Chen *et al* 2006]. In figure 3 above we show some examples of a face image processed by different normalization algorithms: (a) shows the unprocessed image with (b) the original luminance histrogram; (c) is the same image normalized with simple HE and (d) the effect of HE on the image histogram; (e) is the image with *adaptive* HE applied and (f) the effect of AHE on the histogram, in particular note the high frequency blow-up of the histogram; finally (g) shows how CLAHE eliminates the high-frequency artifacts of AHE and (h) reduces the high-frequency blow-up when compared with (f).

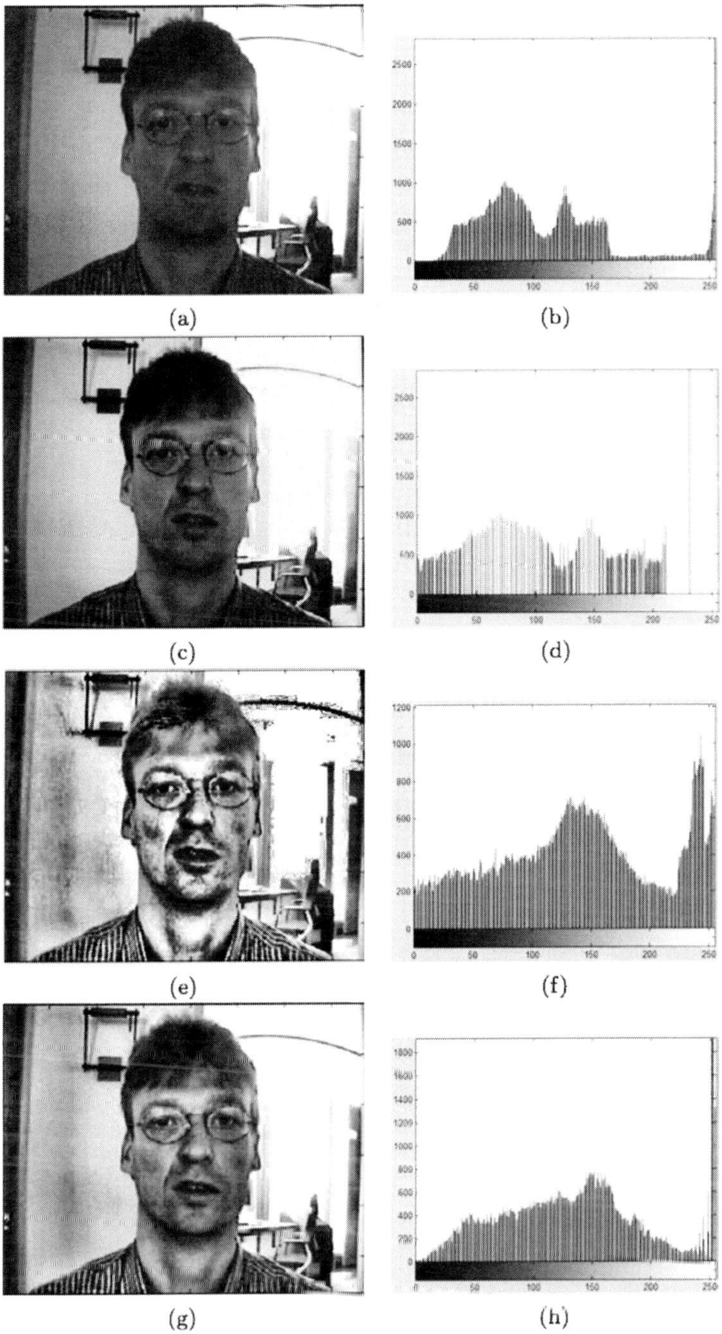

Figure 3. Examples of illumination normalization techniques – details in the text.

5. FEATURE EXTRACTION

Feature extraction for both 1D and 2D HMMs was originally described by [Samaria 1994]. His method was subsequently adopted in the majority of HMM-based face recognition papers. This feature extraction technique is based on scanning the image with a fixed-size window from left-to-right and top-to-bottom. A window of dimensions $h \times w$ pixels begins scanning each extracted face region from the left top corner sub-dividing the image into a set number of $h \times w$ sized blocks.

On each of these blocks a transformation is applied to extract the characterizing features which represent the observation vector for that particular region. Then the scanning window moves towards right with a step-size of n pixels allowing an overlap of o pixels, where o = w − n. Again features are extracted from the new block. The process continues until the scanning window reaches the right margin of the image. When the scanning window reaches the right margin for the first row of scanned blocks, it moves back to the left margin and down with m pixels allowing an overlap of v pixels vertically. The horizontal scanning process is resumed and a second row of blocks results, and from each of these blocks an observation vector is extracted. The scanning process and extraction of blocks is depicted in Figure 4.

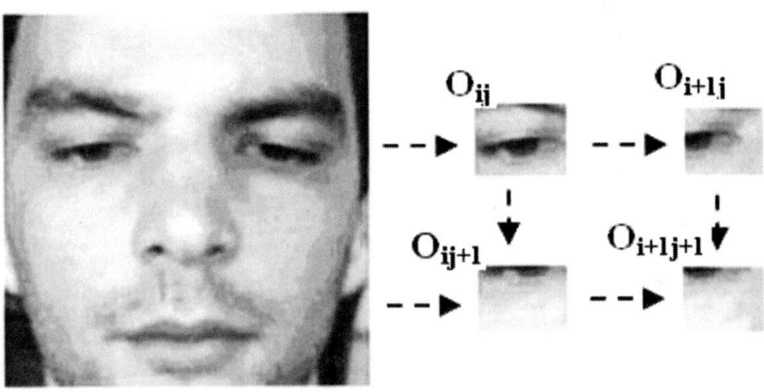

Figure 4. Blocks extraction from a face image

In our work we have used two types of features to describe the images: 2D DCT coefficients and Daubechies wavelets.

5.1. Overview Of Features Used With Hmm In Face Recognition

The first features used in face recognition performed with HMM were pixel intensities [Samaria & Fallside 1993, Samaria 1994, Samaria & Harter 1994]. The recognition rates obtained by Samaria using pixel intensities with a P2D-HMM were up to 94.5% on the ORL database. However the use of pixel intensities as features has some disadvantages [Nefian & Hayes 1999]: firstly they cannot be regarded as robust features since: (i) the intensity of a pixel is very sensitive to the presence of noise in the image or to illumination changes; (ii) the use of all the pixels in the image is computationally complex and time consuming; and (iii) using all image pixels does not eliminate any redundant information and is thus a very inefficient form of feature. Another example of features used with EHMM for face recognition are KLT features used by [Nefian & Hayes 1998, Nefian & Hayes 2000] with recognition rates of up to 98% on ORL database. The main advantage of using KLT features instead of pixel intensities is their capacity to reduce redundant information in an image. The disadvantage is their dependence of the database of training images from which they are derived [Costache 2009].

The most widely used features for HMM in face recognition are 2D-DCT coefficients. These DCT coefficients combine excellent decorrelation properties with energy compaction. Indeed, the more correlated the image is, the more energy compaction increases. Thus a relatively small number of DCT coefficients contain the majority of information encapsulated in an image. A second advantage is the speed with which they can be computed since the basis vectors are independent of the database and are often pre-computed and stored in an imaging device as part of the JPEG image compression standard. Recognition rates obtained when using 2D DCT with HMM can achieve 100% success on smaller databases such as ORL. In our research we also introduce the use of Daubechies wavelets. Apart from the work of [Le & Li 2004] wavelets have not been previously used with HMMs for face recognition applications.

6. FACE RECOGNTION

In the earlier sections of this chapter we have described the main pre-filtering blocks for our AFR system. We nest focus on the actual HMM itself and the various operational processes required to implement the training and recognition phases of our AFR.

6.1. Background to Embedded Hidden Markov Models In Face Recognition

After their introduction in the late 60's by [Baum *et al* 1966, 1970] and the more detailed description in the late 80's by [Raibner & Juang 1986, Rabiner 1989] HMMs have been widely used in speech recognition applications. In this field of application very high recognition rates are obtained due to the specific capacity of HMM to cope with variations in the timing and duration human speech patterns [Juang and Rabiner 2005]. HMMs have also been used successfully in other applications such as OCR and handwriting recognition. Thus it was no surprise that researchers began to consider their use for problems such as face recognition where adaptability of HMMs might offer solutions to some of the underlying problems of accurately recognizing a 2D face region.

Note that the application of HMM techniques to the face recognition problem implies the use of an inherently 1D method of pattern matching to solve an inherently 2D problem. So why did researchers think this might work? Well, as the most significant facial features of a frontal face image occur in a natural order, from top to bottom, and this sequence is immutable, even if the face is moderately rotated. The first attempts to use HMMs for face recognition and detection were made by [Samaria & Fallside 1993, Samaria & Harter 1994] who used a left-to-right HMM and divided the face in a fixed succession of regions (observantion states) such as eyes, nose, & mouth. This early work by Samaria was essentially 1D in scope and the first attempt to implement a more appropriate 2D models was*Pseudo 2D HMM*, introduced by [Agazzi 1994] for character recognition, subsequently adapted by [Samaria 1994] for the face recognition problem. The idea was later taken forward and improved by [Nefian & Hayes 1999, 2000]. These researchers changed the name to *Embedded HMM* (EHMM).

There have been several alternative 2D versions of HMM used for face recognition in the literature. However EHMM has become the standard method employed by researchers working in the field of HMM face recognition. As a result this algorithm has been implemented in the Intel digital image processing C++ library [OpenCV 2006] which was also employed to implement our face detector, described in section 2 above.

6.2. An Overview Of Ehmms

The embedded HMM is a generalization of the classic HMM, where each state in the one dimensional HMM is itself an HMM. This enables a generalization of the 1D HMM techniques to a second dimension while simplifying the dependencies and transitions between states. Thus, an embedded HMM consists of a set of super states each of which envelopes a set of embedded states. The super states model the two dimensional data in a first dimension, while the embedded HMMs model the data in the other dimension.

The structure of an EHMM with 3 superstates and 4 embedded states is shown in Figure 5(a). This EHMM is unrestricted, meaning that transitions between all the states in the embedded HMMs and between all the superstates are allowed.

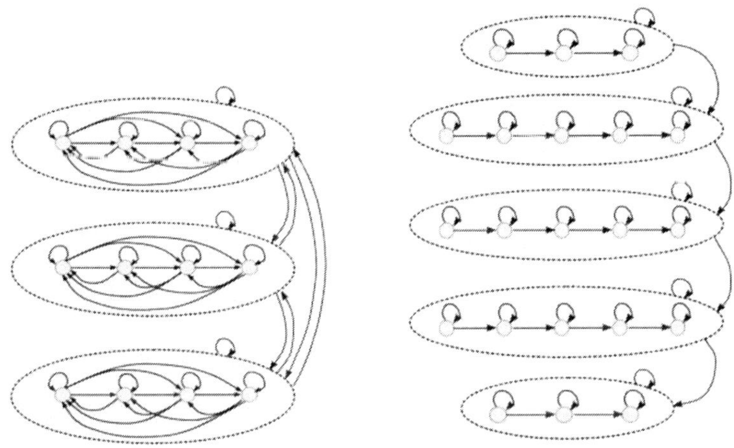

(a): An unrestricted EHMM (b): Restricted EHMM for face recognition

Figure 5. EHMM for face recognition

The elements of an embedded HMM are:

1 A set of N_0 superstates $S_0 = S_{0,1}$, $1 \le i \le N_0$,

2 The initial probabilities of the super states Π0=π0,iΠ0=π0,I where

3 π0,iπ0,I is the probability of being in superstate i at time zero.

4 The transition probability matrix A0 = a0,ij, where a0,ij is the probability of transitioning from super state i to superstate j.

5 The parameters of the embedded HMM for the superstate k, 1≤k≤N0 and where $\Lambda^k = (\Pi_1^k, A_1^k, B^k)$ which includes: the number of embedded

states in the kth super state, N_1^k, and the set of embedded states, $S_1^k = S_{1,i}^k$ with $1 \le i \le N_1^k$; the initial state distribution, $\Pi_1^k = \pi_{1,i}^k$, where $\pi_{1,i}^k$ is the probability of being in state i of the superstate k at time zero; the state transition probability matrix $A_1^k = a_{1,ij}^k$, where $a_{1,ij}^k$ is the transition probability from state i to state j; the probability distribution matrix of the observations, B^k ; these observations are characterized by a set of continuous probability density functions, considered finite *Gaussian mixtures* of the form:

$$b_i^k(O_{t0,t1}) = \sum_{m=1}^{M_i^k} c_{im}^k N(O_{t0,t1}, \mu_{im}^k, U_{im}^k) \tag{2}$$

where

c_{im}^k is the mixture coefficient for the mth mixture in state i of super state k, and $N(O_{t0,t1}, \mu_{im}^k, U_{im}^k)$ is a Gaussian density with a mean vector Mkimμimk and covariancematrix U_{im}^k.

In shorthand notation, an embedded HMM is defined as the triplet $\lambda = (\Pi_0, A_0, \Lambda)$ where $\Lambda = \Lambda^1, \Lambda^2, \ldots, \Lambda^{N_0}$. This model is appropriate for facial images since it exploits an important characteristic of these: frontal faces preserve the structure of "superstates" from top to bottom, and also the left-to-right structure of 'embedded states' within each "superstate" [Nefian & Hayes 1999,2000]. An example of the state structure of the face model and the non-zero transition probabilities of the embedded HMM are shown in Figure 5(b). The configuration presented is 5 super states, each with 3, 6, 6, 6, 3 states respectively. Each state in the overall top-to-bottom HMM is assigned to a left-to-right 1D HMM. In this case transitions are allowed only from left-to-right or self-transitions and only between *consecutive states* both for the embedded HMMs within each superstate, and for the main superstates of the top-level HMM as well.

6.3. The Training Process For An Ehmms

The training of HMM, as shown by [Rabiner 1989] is accomplished using the Baum-Welch algorithm. While EHMM exhibits a more complex structure than the simple 1D HMM, the training algorithm follows the same steps. The main difference in training is the use of a doubly embedded Viterbi for segmentation. The training structure is depicted in Figure 6, and the role of each block is described next:

Step 1. PrototypeHMM: the first step is defining the prototype EHMM: *parameters:* Nki the numbers of states, N0 the number of superstates, *K* the number of Gaussians used to model the probability density for the observation vectors in each state of an embedded HMM; *conditions:* which transitions are allowed $a^k_{1,ij} > 0$) and which are not ($a^k_{1,ij} = 0$); in our left-to-right HMM the only transitions allowed are self-transitions and transitions to the next state, so the probability of transition to previous states is 0. For a numerical example we choose $N_0 = 5$, $N^k_1 = 3,6,6,6,3$ where *k* = 1, 2,..., 5, and *K* = 3.

Step 2. Uniform segmentation: the image is uniformly segmented. Firstly the observation vectors extracted from the entire image are uniformly divided into N−05 vertical super states, or image strips, for the overall top-to-bottom HMM. Next the data corresponding to each of these vertical super states is now horizontally segmented from left to right into N^k_i uniform states. For a 128 × 128 pixel facial region with scanning window 12 × 12 with 8 pixels overlap we obtain 30 observation vectors per scanning row both horizontally and vertically, thus 900 observation vectors in total. In a uniform segmentation, the observation vectors are first divided between N=05superstates: 30 observation vectors per row/5 superstates, so 6 observation vectors per row in each superstate ⍰ 6 × 30 = 180 observation vectors per superstate. Then horizontally these 180 observation vectors are uniformly divided in states as follows: for the superstates 1 and 5 with only 3 states: there will be 60 observation vectors per state; for superstates 2, 3, 4 with 6 states each: there will be 30 observation vectors per state.

Step 3. Parameter initialization: after segmentation, the initial estimates of the model parameters are obtained using the concept of counting event occurrences for the initial probabilities of the states and the transition probabilities. In order to compute the observation probabilities, for each state of the embedded HMMs a K-means clustering algorithm, where *K* is the number of Gaussians per state, is applied. All the observation vectors extracted from each state are used to obtain a corresponding mixture of Gaussians describing the observation probability density function. From these initial values we then begin to iterate. In the example given above the initial probabilities of the states in each superstate are determined from the system constraints as follows: first state in each embedded HMM has initial probability equal to 1.0, all the other states have initial probability of zero. Transition probabilities are then obtained by counting transition occurrences. For example in the first state of the first superstate: there are 60 observation vectors distributed across 6 horizontal rows of scanning implying 6 possibilities of transition from state 1 into state 2: probability of transition from state 1 into state 1 is $P_{1,1}$=54/60$P_{1,1}$=54/60; probability of transition from state 1 into state 2 is $P_{1,2}$=6/60$P_{1,2}$=6/60 ; transition probabilities for the other states can be calculated in the same way.

Step 4. Embedded Viterbi segmentation: in the first step of the iteration, a doubly embedded Viterbi algorithm replaces the uniform segmentation. With the new segmentation and again counting event occurrences, a set of new values for initial and transition probabilities are found. This process is described in detail in section 5.4 below.

Step 5. Segmental K-means: according to the new segmentation performed at step 4, another K-means is applied to the current set of observation vectors corresponding to each new state, and new observation probability density functions are computed. On the next iteration, these new values are introduced into the doubly embedded Viterbi and a new segmentation is initiated.

Step 6. Convergence: Steps 4 and 5 are repeated until the difference on consecutive iterations is below a set threshold. If convergence is not achieved after a certain number of iterations the training is considered to have failed for the current input face region. Typically we have set the convergence threshold at 0.01 and the maximum number of iterations at 40. Once convergence is achieved, further iterations are stopped and the EHMM is output and stored in a reference database.

6.4. The Decoding Process For An Ehmm (Doubly Embedded Viterbi)

In the description of the training process above we have seen that step 4 consists in the re-segmentation of the states in the 1D HMMs and of the superstates in the overall HMM. Re-segmentation means finding the most probable sequence of states given a certain sequence of observation vectors and we can solve this problem by applying the Viterbi algorithm. We can easily apply Viterbi algorithm in the embedded 1D HMMs for which we have determined all the probabilities at step 3 above. However for the overall HMM after step 3 we only have the initial and transition probabilities, without the observations probabilities.

In order to solve this problem a method based on the Viterbi algorithm known as *double embedded Viterbi* was developed [Agazzi 1994]. It involves applying the Viterbi algorithm to both the embedded HMMs and to the global, or top-level HMM, hence the name. A detailed description may be found in [Nefian 1999]. As the algorithm is mathematically complex and the formulas are challenging to understand and even more so to implement. For this reason we will next provide a detailed practical (as opposed to theoretical) description of our step by step implementation of the algorithm. The underlying concept is illustrated in Figure 6 and comprises the following steps:

Step 1. After the parameters initialization step No. 3 of the previous section we have: initial probabilities, transition probabilities and observation probabilities for each embedded HMM, and initial and transition probabilities for the top-level HMM. In the first step of the double Viterbi, each scanned row of observation vectors $O_0^i = v_j^i$ with $1 \le i \le (H-v)/(h-v)$ and $1 \le j \le (W-o)/(w-o)$ within each of the embedded 1D HMMs has the conventional Viterbi algorithm applied. After this step the optimal state distribution is obtained for each row of observation vectors based on the relevant 1D HMM $(Q^i | Q_0^i, \Lambda_k)$ and also the probability of each row of observation for the top-level HMM, or superstate $P(O_0^i, Q^i | \Lambda^k)$, where $1 \le i \le (H-v)/(h-v)$ and $1 \le k \le N_0$.

Step 2. After the first application of the Viterbi algorithm we have: initial and transition probabilities for the superstates as determined at step 1 of the training algorithm described above, and the observations probability distributions for the top-level HMM, that is: the probabilities of each horizontal row of observations given each superstate. Now Viterbi is applied on the top-level HMM and the optimal sequence of superstates is obtained given the sequence of rows of observation vectors (vertical re-segmentation) and the probability of the entire sequence of observations (which characterizes the entire image) given the EHMM model created. This probability is compared at each iteration with the same probability obtained on the previous iteration (step 6 in Section 5.3).

Step 3. Vertical re-segmentation means reassignment of each row of observation vectors to the corresponding superstate (embedded 1D HMM). After we determine to which embedded 1D HMM each row of observation vectors appertains. Then using the findings at the first step of the*double embedded* Viterbi algorithm horizontal re-segmentation can be achieved.

In Figure 7 below we give an example of states and superstates segmentation for a given face image. Each color represents a different state in the superstates. As one can see, the 5 superstates found in this image are:

forehead region, ends right above the eyebrows and is divided into 3 states, eye region, starts just above eyebrows and ends just after the pupil, is divided into 6 states, the nose region, starts just after the pupil and ends just before the nostrils, is divided into 6 states, the nostrils region, starts just under the nose region and ends at the middle between mouth and tip of the nose, is divided into 6 states, and finally the mouth region starts after the nostrils region and ends at the bottom of the image, is divided into 3 states. Also from the image we can see that the rows of observation vectors inside one superstate are distributed unevenly between states.

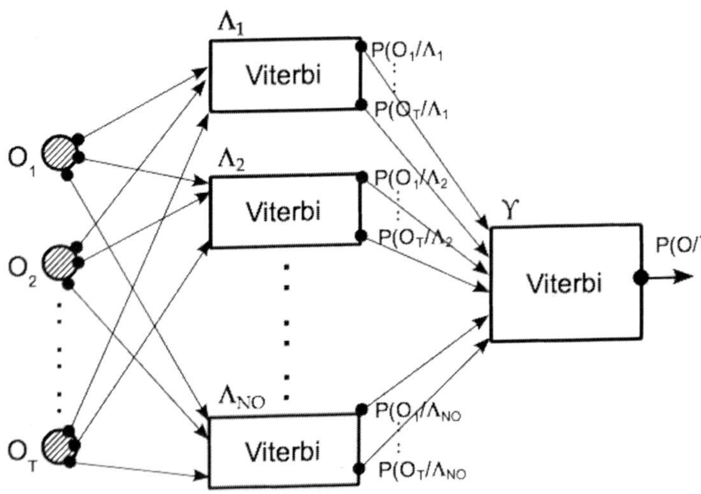

Figure 6. Doubly embedded Viterbi

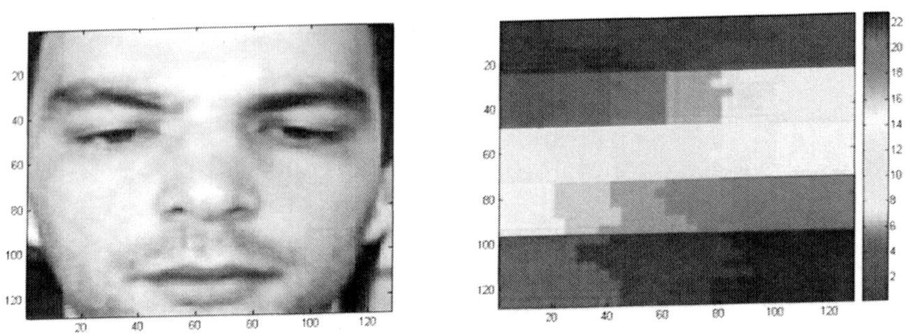

Figure 7. State distribution after applying doubly embedded Viterbi

6.5. The Evaluation Process For An Ehmm

In the training process described previously we have shown how an EHMM model is built for a subject in the database. After building separate EHMMs for all subjects in the database we can move to the recognition step where the likelihood of a face test image is evaluated against all models. The evaluation process comprises of the following steps:

Step 1. first the face image is scanned with a rectangular window from left-to-right and top-to-bottom and the observation vectors are extracted.

Step 2. then the sequence of observation vectors is introduced in each EHMM model and its corresponding likelihood is computed. Theoretically the probability of a sequence of observation vectors, given a model, is found using the forward-backward *evaluation algorithm*. However, in practice it was argued [Raibner 1989, Agazzi 1994] that the Viterbi algorithm can successfully replace the evaluation algorithm. For our EHMM we use a doubly embedded Viterbi algorithm. At the end of this step we have the probabilities of the test image to match each of the EHMM models in the recognition database.

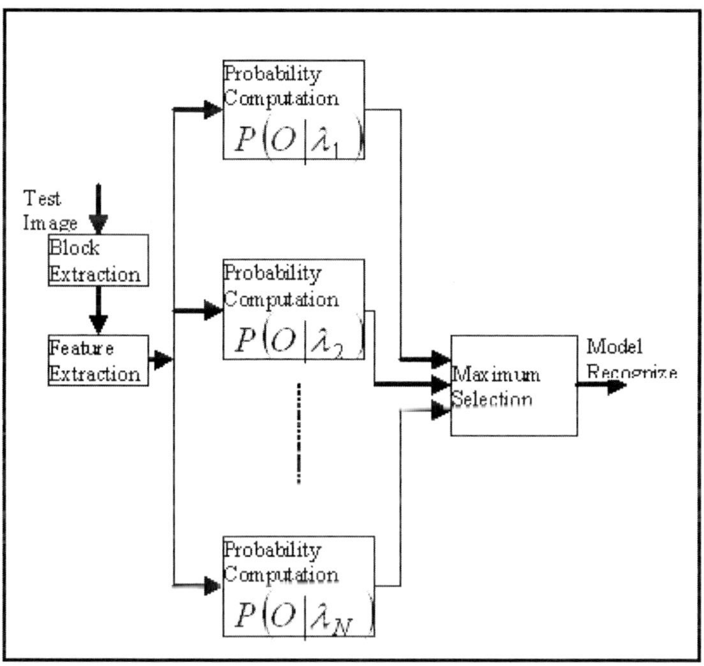

Figure 8. HMM recognition scheme (N is the number of subjects in the database)

Step 3. the final step consists in comparing all the probabilities computed at the previous step and choosing as winner the model which returns the highest probability. The evaluation process is depicted graphically in Figure 8.

7. IMPLEMENTATION DETAILS

In order to implement our AFR system two different software programs were designed: one for the face detection and normalization processes and one to support the HMM based face recognition process. Many functions for face detection and recognition were based on a well known open source image processing library [OpenCV 2006]. Some details on each of these software workflows are given to facilitate other researchers.

7.1. Face Detection

For the detection and cropping of all faces in the test databases we employed a well-known face detection algorithm [Viola & Jones 2000, 2001], described in section 2 above. In order to implement detection and cropping of all faces in all images in a single step, a tool was required to operate batch processes. This is implemented using Matlab. Such an approach allows additional high-level filters and other image processing techniques, also implemented in Matlab, to be easily linked with the OpenCV based face detection process. Thus the speed and efficiency of the OpenCV routines are coupled with the flexibility to incorporate supplemental Matlab filters into our test and evaluation process.

The Matlab program takes as input a folder of images, automatically loading each image, calling the face detection function from OpenCV and returning all face rectangles detected in the given images. These are then cropped and saved on disk as new JPG image files. Note that this process facilitates a manual inspection or supplemental testing of a set of images to determine if they are correctly and uniformly cropped. To achieve the functionality of this program, there are two principal stages:

- the declaration of the face detection function in OpenCV has to be modified in order to provide Matlab access to the detection function from the detection library using the standard mex (Matlab-C) interface[1] - ;

- the OpenCV library is compiled with the Matlab-C compiler. The mex command was used to build the detection function adding all OpenCV dependencies.

At the end of this stage all detected faces were manually separated by subject in different folders. Throughout our tests we used different numbers of pictures per person in the training stage, and a fixed number of pictures in the testing stage. More exactly, we trained the system successively with 1 and up to 5 pictures per person, and we tested with the remaining 5 pictures that were not used in any training. In future tests we will denote the number of faces used for training as Nvs5: 1vs5, 2vs5, etc.

7.2. Face Recognition

The second step in implementing the face recognition system was to build a program that would perform the main face recognition processes. The face recognition implementation was done in the C language using Microsoft Visual Studio. The implementation consists of three main components:

1 Top-level component is the first component and has the purpose of reading multiple images from the disc,saving the output of the training stage (which is represented by the models) and analyzing the output of the testing stage. A detailed description of the training and testing stages can be found in Chap. 4;

2 Mid-level component: the second component which processes (pre-processing: illumination normalization, resize, filtering etc) the faces, computes observation vectors, builds and stores HMM models and computes likelihoods between faces and models. Please see Chapter 4 for a description of the HMM stages;

3 Low-level component is the third component which contains the basic routines of the EHMM algorithm (feature extraction, segmentation, Viterbi, state probability distribution etc) and uses functions implemented in the OpenCV library.

A detailed description of the processes taking place in each of these components and how they are interfaced is given in [Iancu 2010]. Figure 9 presents a visual explanation.

7.3. Databases and Training Datasets

For the experiments presented in the next sections of this chapter we used a mix of subjects from 3 standard databases: BioID, Achermann and UMIST. Each database provides some of our desired variations: BioID exhibits high variations in illumination, some expression variations and slight pose variations; Achermann presents some head rotations and slight illumination variations; UMIST covers a range of poses from frontal to semi-profile. A short description of each of these databases is given next.

BioID database: BioID[2] - is a dataset consisting of 1521 gray level images with a resolution of 384 × 286 pixels, containing 23 different test persons with frontal views with variations in facial expression and illumination. The actual size of the face inside the picture is on average 128 × 128. From the entire database 200 pictures of 20 different persons were selected. Faces were selected to maximize pose and illumination variations as far as possible in the selected picture.

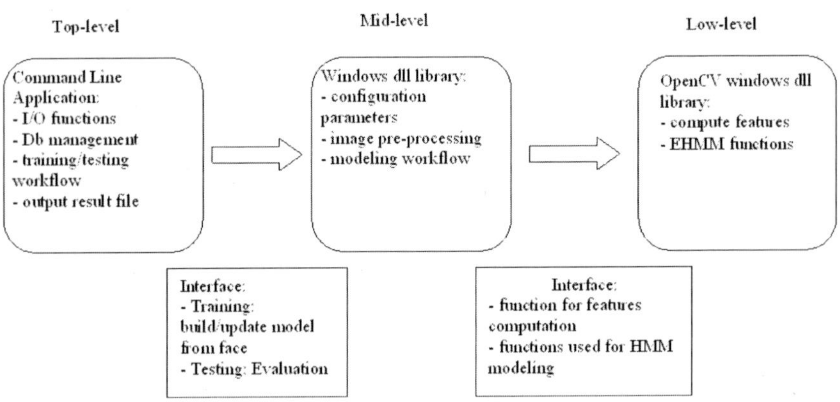

Figure 9. Main program workflow

Achermann database: The Achermann database[3] - contains 260 images of 26 people, each with 10 images, with 143×143 pixel in size. Unlike many other databases which contain only frontal views, the faces in this database span a range of pose angles.

UMIST database: UMIST database[4] - consists of 564 images of 20 people, each covering a range of poses from profile to frontal views. Subjects cover a range of race/- sex/appearance. The files are all in PGM format, approximately 220 × 220 pixels in 256 shades of grey. From this database we extracted 100 pictures of 10 subjects, with pictures ranging from frontal up to 30 degrees right and left turn.

Tests were performed using the software tools described in section previously in Section 6.2 on a database formed by combining the 3 databases described above. This comvined database consisted of 560 pictures of 56 subjects, for each subject 10 pictures being selected. Unless stated otherwise, pictures were resized to 128 × 128 grayscale pixels then each face was scanned using a 12 × 12 window from left to right and top to bottom, with an overlap of 8 pixel both vertically and horizontally, in order to extract the observation vectors. The first 3×3 2D DCT coefficients corresponding to the low frequencies are retained. The EHMM depicted in Figure 5(b) is used, with 5 super states and 3-6-6-6-3 embedded states respectively, representing forehead-eyes-nose-mouth-chin areas. A standard number of 3 Gaussians were used in each state to model the feature vectors for most of the following experiments. In the training step, 1 to 5 pictures for each subject are used for building the corresponding EHMM model. After training, in order to check the algorithm, *verification* is performed, where the images used for training are also used for testing. If the algorithm is correct than the recognition rate should be 100%. In the recognition phase, the 5 images per person not used for training are utilized.

8. EXPERIMENTAL RESULTS

As structure for the EHMM we used Samaria's thesis as starting point. According to Samaria's experiments presented in his PhD thesis [Samaria 1994], the most efficient structure allowing the smallest error rate (5.5%) consists in 5 super states with 3-6-6-6-3 embedded states respectively and was tested employing as observation vectors the pixel intensities of a face region and the dataset of face regions was drawn from the ORL database. The same EHMM structure was used by [Nefian 1999] 2D DCT coefficients as observation vectors and the dataset of face regions was drawn from the ORL database. Nefian obtained an improved error rate of 2%.

8.1. Test Data 1 – Different Sizes Of Face Region

A first set of experiments is directed towards determining an optimal size of the detected and cropped face regions used for recognition. The sizes of the faces available in research databases can vary significantly. As examples, it varies between 128 × 128 for BioID, 143 × 143 for Achermann and 220 × 220 for UMIST. As our experiments draw images from these, and other data sources, a consistent normalization between face regions is important.

Downscaling will result in information loss, but this is still preferable to upscaling which can introduce artifacts and create false information. There is also a reduction in computational demands for smaller pictures – another reason to favour downscaling. For example it takes 75% less time to process a 64×64 image region than for 128 × 128 sized regons. Thus processing smaller pictures is clearly desirable for algorithms to run efficiently in handheld devices such as digital cameras. In this set of experiments we tested sizes between 64 × 64 and 196 × 196. The bicubic spline interpolation technique was used for image downscaling where required. The results are given in the Table 1 and depicted in Figure 10.

Table 1. Recognition rates (%) for different picture sizes

Picture size	1vs5	2vs5	3vs5	4vs5	5vs5
64 × 64	42.86	56.79	73.57	83.57	89.64
96 × 96	52.14	64.64	76.43	86.07	87.86
128 × 128	66.43	71.07	81.07	83.57	86.07
192 × 192	61.79	69.29	77.86	83.93	83.57

8.2. Test Data 2 – Different Numbers Of Gaussians

One of the underlying assumptions of our HMM model is that a random signal can be modeled by one or more Gaussian probability functions. If the signal is simple, it clusters around one mean value and it can be modeled by a single Gaussian. For more complex signals there may be more than one clustering center, and a more accurate model will match the number of Gaussians to the number of primary clusters. The objective of this experiment is to test how a varying number of Gaussians will influence the underlying recognition rates. We used face regions of size 128 × 128 pixels. This size of images should provide close to an optimal recognition rate as detailed in section 8.1.

The best results are obtained for mixtures of 3 Gaussians per HMM. For 2 or 4 Gaussians per HMM good results are also obtained, particularly when more training images per person are available. Note, however, that 4 Gaussians per HMM requires significantly more computation time and the benefits are marginal. Because the difference in processing time between modeling with 2 and 3 Gaussians is not significant (the recognition process with 2 Gaussians is in average 15% faster than for 3 Gaussians), it was decided to use a mixture of 3 Gaussians for our later experiments.

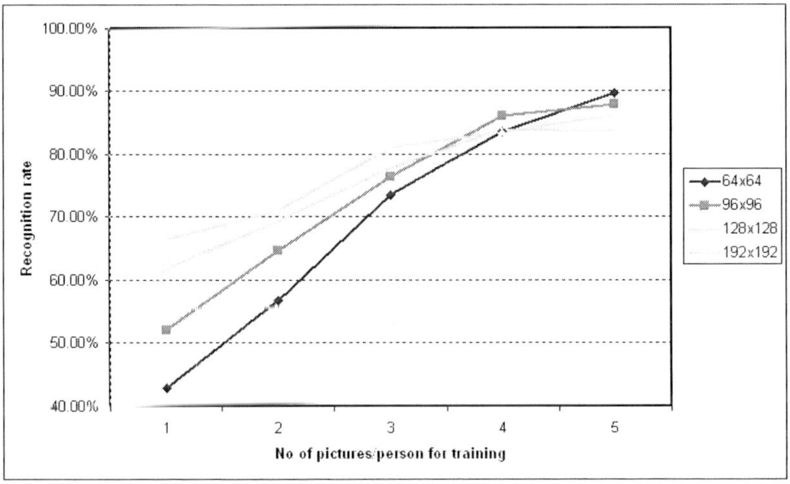

Figure 10. Recognition rates for different picture sizes

Table 2. Recognition rates (%) for different numbers of Gaussians

Picture size	1vs5	2vs5	3vs5	4vs5	5vs5
64 × 64	42.86	56.79	73.57	83.57	89.64
96 × 96	52.14	64.64	76.43	86.07	87.86
128 × 128	66.43	71.07	81.07	83.57	86.07
192 × 192	61.79	69.29	77.86	83.93	83.57

Note that for more complex or unusual model characteristics it is expected that the improvement in general recognition rates for 3 Gaussians could be more significant than is shown here. For now we can state that increased computation from using 3 Gaussians is likely to be balanced by the adaptability that is available to the model to handle more extreme variations in facial characteristics. No investigation of varying the underlying number of Gaussians for each superstate was undertaken. However it seems likely that superstates which cover complex features, such as the eye, nose and mouth regions are likely to benefit from increasing the underlying number of Gaussians in the more complex face regions.

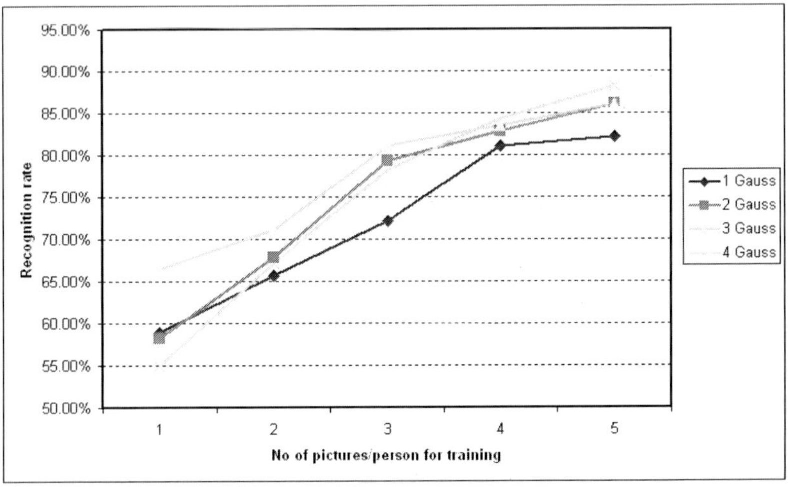

Figure 11. Recognition rates for different numbers of Gaussian functions

8.3. Experimental Results For Different Numbers Of Dct Coefficients

Again, the idea behind using fewer features to represent a signal is to speed up the recognition process, without adversely affecting the recognition rates. The number of 2D DCT coefficients we used previously was 3 × 3. In Figure 12 we show the results for 2×2, 3×3 and 5×5 DCT coefficients used. The results are more clearly presented in Table 3. We can see that on average 3×3 coefficients perform marginally better than 5 × 5 when up to 3 training pictures are used. However as more pictures are used in training, 5×5 coefficients seem to characterize the image better although in many applications this improvement is may not be sufficient to justify the additional computational effort for handheld imaging devices.

Table 3. Recognition rates (%) for different numbers of DCT features

No. of Gaussians	1vs5	2vs5	3vs5	4vs5	5vs5
1	58.93	65.71	72.14	81.07	82.14
2	58.21	67.86	79.29	82.86	86.07
3	66.43	71.07	81.07	83.57	86.07
4	55.00	67.14	78.21	84.29	88.21

8.4. Experimental Results For Simplified Topologies

The final tests of this section involve using a different topology for the EHMM. The core model was drawn from the work of Samaria and Nefian employs 5 superstates with an internal organization of 3-6-6-6-3 embedded states respectively, on images resized at 128 × 128. Here a simplified version of the classic EHMM is tested: the face is segmented into 4 superstates and 2-4-4-2 embedded states. Smaller face regions are used to further reduce the memory and computation required. Face regions were resized to 32 × 32 and 64 × 64. Details of the tests and results are presented in Figures 13& 14. The numerical results can be found in Tables 5 and 6.

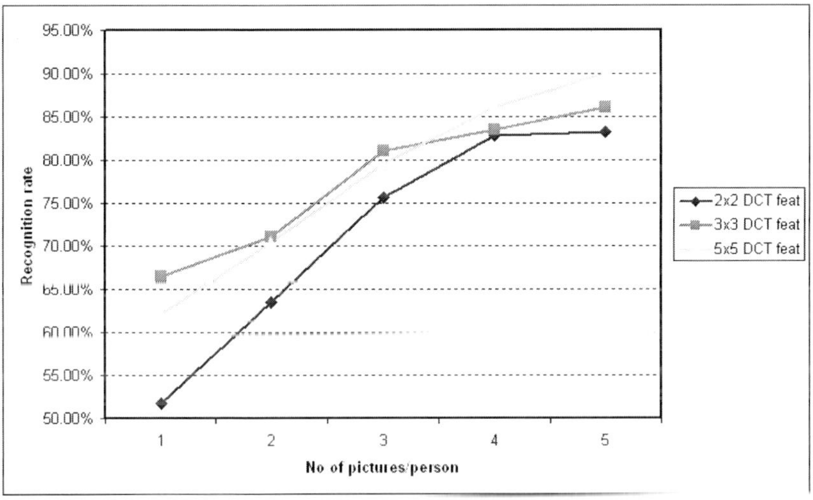

Figure 12. Recognition rates for different numbers of features

Table 4. Recognition rates (%) for picture size 32 × 32

Window size / No Gaussian	1vs5	2vs5	3vs5	4vs5	5vs5
8 × 8/1 Gauss	57.86	67.50	72.86	80.00	82.14
12 × 12/1 Gauss	51.43	51.79	64.29	76.43	83.21
8 × 8/2 Gauss	58.57	64.64	76.43	84.29	86.43
12 × 12/2 Gauss	55.36	49.64	61.07	74.64	82.14

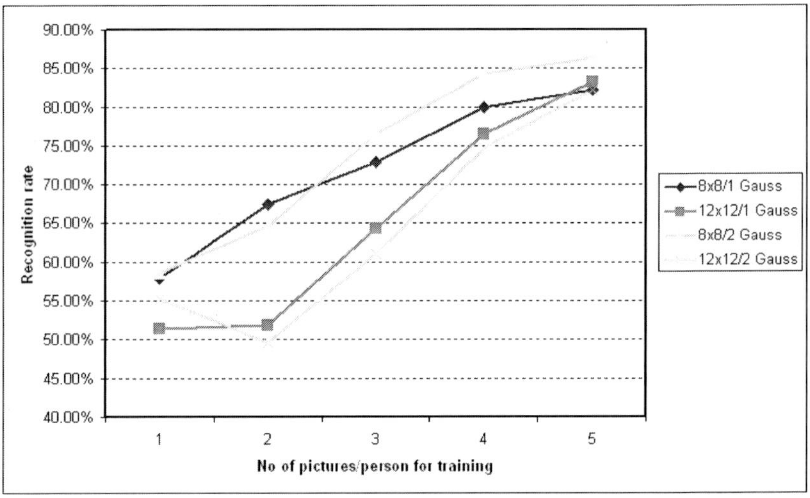

Figure 13. Recognition rates for 32 × 32 picture size

There were other parameters we had to change in the tests in order to increase the recognition rates. Originally we started the tests on 32 × 32 pixels images scanned with a 12 × 12 pixels window with 8 pixels overlap (75% of the size of the scanning window). However a simple calculation shows that in this case we only have 6 scanning steps both vertically and horizontally so there will be superstates with only 6 observation vectors, that meaning that some states will have only 1 or 2 observation vectors to be modeled with Gaussian mixtures. Obviously modeling with 2 Gaussian mixtures a single observation vector does not make sense. Furthermore having a single observation vector per state cannot offer a detailed description of that state.

Because of this, low recognition rates were obtained for 32 × 32 pixels images scanned with 12×12 pixels windows. However, for images sized 64 × 64 pixels, scanning with an 8 × 8 pixels window could mean taking too much detail and missing larger facial features. This may explain why the best results for 64 × 64 size images are obtained when scanning with 12 × 12 pixels window.

Table 5. Recognition rates (%) for picture size 64 × 64

Window size/No. of Gaussians	1vs5	2vs5	3vs5	4vs5	5vs5
8 × 8/1 Gauss	55.71	61.79	69.64	75.36	76.43
12 × 12/1 Gauss	60.71	65.36	72.50	78.57	82.50
8 × 8/2 Gauss	59.64	63.93	71.79	79.64	82.86
12 × 12/2 Gauss	57.14	65.71	70.36	81.07	85.00

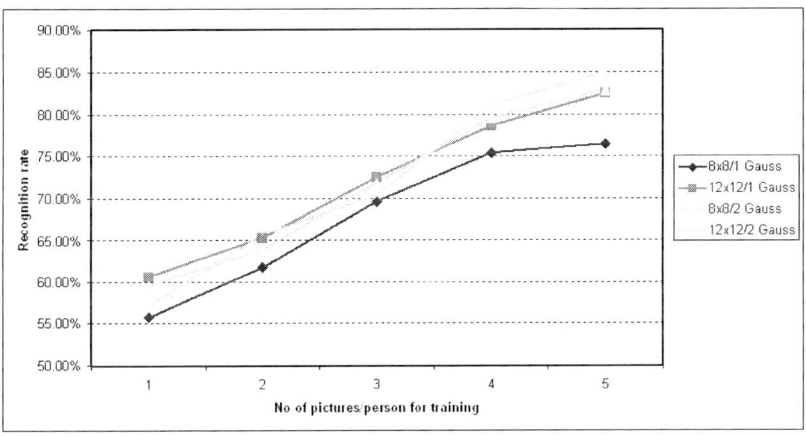

Figure 14. Recognition rates for 64 × 64 picture size

8.5. Experiments Using Different Illumination Normalization Techniques

For our experiments to tackle the problem of illumination variations a study was undertaken of the effectiveness of the following three illumination normalization algorithms:

1 *histogram equalization* (HE) based on [Gonzalez & Woods 1992],

2 *contrast limited adaptive histogram equalization* (CLAHE) based on [Zuiderveld 1994], and

3 the relative new method of DCT in the logarithm domain - *logDCT* based on [Chen *et al* 2006].

In addition paired combinations of several of these techniques were also evaluated. To the best of our knowledge this is the first attempt in the literature to compare different illumination normalization techniques in the context of EHMM.

Table 6. Recognition rates(%) for illumination normalization techniques

Illum Norm Tech	1vs5	2vs5	3vs5	4vs5	5vs5
No normalization	66.43	71.07	81.07	83.57	86.07
HE	65.00	74.64	85.71	90.71	92.50
CLAHE	68.21	75.71	82.86	88.21	90.57
LogDCT	52.50	58.21	70.71	76.43	77.86
LogDCT+CLAHE	60.00	70.36	75.00	82.14	84.64
CLAHE+LogDCT	68.21	77.50	84.64	90.00	92.86
LogDCT+HE	66.43	72.86	81.43	86.79	90.00
HE+LogDCT	53.57	64.29	76.79	85.00	88.57
HE+CLAHE	68.57	78.21	86.07	92.14	95.36
CLAHE+HE	71.07	79.29	87.14	92.86	95.71

We tested these three illumination normalization techniques, *viz*, HE, CLAHE, LogDCT and compared the recognition results with the results obtained when no illumination normalization was performed. After visually observing the results of illumination normalization on each image in the database we concluded that LogDCT flattens the facial features excessively whereas CLAHE enhances them. Thus the idea of combining these two normalization methods, was mooted. It was then decided to investigate a number of other combinations of these techniques and the results are presented below. The number of DCT components that are canceled for logDCT depends on the size of the image and the level of the illumination variation present in the image. For our tests we used the first 4 DCT coefficients for 128 × 128 images. The recognition rates for each illumination normalization technique and the combinations we tested are given in Table 6.

In Figure 15 a set of face images affected by various degrees of illumination are shown together with the results of applying each illumination normalization technique or combinations thereof to the original images. Making a parallel between the recognition rates given in Table 6 and the visual results shown in Figure 15, a few observations can be provided:

– the best recognition rates are obtained when combining HE and CLAHE regardless of the order.

– Better recognition rates are achieved using the simple HE technique when compared to the more sophisticated CLAHE and LogDCT techniques.

– LogDCT returns very poor results even when compared to the rates obtained when no illumination normalization technique is used.

– From a human's perceptual point of view, the most efficient visually illumina- tion normalization technique appears to be LogDCT+CLAHE but this gives recognition rates poorer than the original pictures

– The best recognition rates are obtained for illumination normalization which enhances the facial features: HE, CLAHE and combinations of these two.

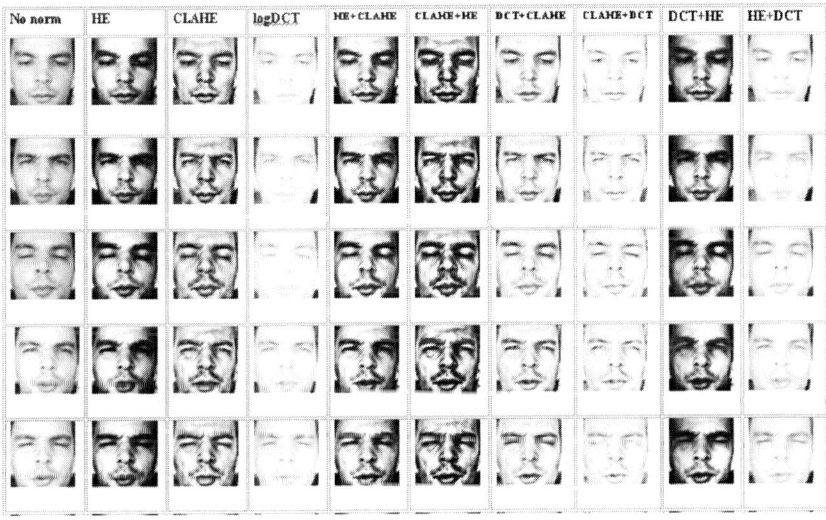

Figure 15. Illustrative examples of various illumination normalization techniques applied to a representative subset of face data.

There are two exceptions that contradict this last rule: (i)LogDCT+HE: despite the fact that from Figure 15 this combination appears to enhance facial features, it also appears to be unable to eliminate the influence of illumination;(ii)The second exception is CLAHE+LogDCT which gives quite good results despite looking somewhat flatter than the other techniques.

9. DISCUSSION OF OUR EXPERIMENTS

Throughout this chapter we described a series of tests performed with the purpose of finding the optimal combination of factors that influence the recognition process: size of face image, topology of the model, that is number of superstates, number of states for each super state and number of Gaussians to model the observation vectors, illumination normalization technique to diminish the differences in illuminations, and coefficients to describe the information contained in the image.

In order to make things more difficult for our recognizer, we tried to emulate as best possible the less that ideal appearance and diversity of a consumer collection by combining three different standard databases in one big collection of images. The databases used were described briefly in Section 7.3. Also, throughout the tests we used different numbers of faces of the subjects for tests and training: from 1 to 5 faces for training and the remaining 5 faces for testing. From the results presented in Section 8 the following conclusions can be reached:

- An optimal size for face image when using HMM is 128 × 128 pixels, although the recognition rates obtained for the smaller size of 96 × 96 pixels are quite close and may offer a better choice for applications where memory and computational efficiency are important, e.g. in handheld imaging devices;
- From experiments performed in section 8.2 the optimal number of Gaussian functions is 3, representing a trade-off between best precision and computational burden; but we remark that with 2 Gaussians a faster computation is achieved with almost the same recognition performance; the effects of varying the number of Gaussians across super-stages was not considered in this research;

- The optimal performance with 2D DCT is achieved by employing the first 9 coefficients, but we noted in section 8.3 that using the first 4 coefficients gives an acceptable result as well and may be preferable where speed of computation and memory efficiency are important; no significant improvement was noted when we used Daubechies wavelets in place of DCT;

- In section 8.4, very good results were obtained for a reduced 2-4-4-2 EHMM topology applied on very small images: these results improve when increasing the number of training images per person, and recognition rates as high as 86.43% were achieved in our experiments. As no illumination normalization was used and these tests were performed on a combined database rather than a single standard database, the results which were obtained may be regarded as highly promising for real-world applications.

- Finally, in Section 8.5 three different illumination normalization techniques are used in the pre-processing phase of our recognizer. We also investigated some non-standard combinations of these techniques to determine their suitability for pre-processing data for a HMM face recognition algorithm. The best results were obtained for CLAHE and HE with over 95% recognition rates. Very good performances were obtained when using a combination of CLAHE and logDCT, with 92.87% recognition rate, but also when using the more basic HE, with up to 92.5%. This analysis of various image normalization filters should pro- vide a useful baseline for future researchers in face recognition. One aspect we would have liked to investigate was the potential of such combining of illumination normalization filters to improve the performance of other well-known face recognition techniques such as PCA, ICA and AAM methods.

NOTES

1. http://www.mathworks.com/support/compilers/interface.html
2. http://www.bioid.com/support/downloads/software/bioid-face-database.html
3. http://iamwww.unibe.ch/fkiwww/staff/achermann.html
4. http://www.sheffield.ac.uk/eee/research/iel/research/face.html

REFERENCES

1. Y. Adini, Y. Moses, S. Ullman, 1997Face recognition the problem of compensating for changes in illumination direction IEEE Trans. on Pattern Analysis and Machine Intelligence, 7 19 721 732

2. L. E. Baum, T. Petrie, 1966 Statistical inference for probabilistic functions of finite state markov chains. Ann. Math. Stat, 6 37 1554 1563

3. L. E. Baum, T. Petrie, G. Soules, N. Weiss, 1970A maximization technique occurring in the statistic alanalysis of probabilistic functions of markov chains. Ann. Math. Stat, 1 41 164 171

4. R. Chellappa, B. S. Manjunath, C. V. D. Malsburg, 1992A feature based approach to face recognition IEEE Conference on Computer Vision and Pattern Recognition, 373 378 .

5. R. Chellappa, C. Wilson, S. Sirohey, 1995 Human and machine recognition of faces: A survey. Proc. of IEEE, 83 5 705 741

6. W. L. Chen, M. J. Er, S. Q. Wu, 2006 Illumination compensation and normalization for robust face recognition using discrete cosine transform in logarithm domain. Systems, Man and Cybernetics, Part B, 36 458 466

7. P. Corcoran, C. Iancu, G. Costache, 2006Improved HMM based face recognition system. OPTIM, Brasov, Romania, 4 143 146

8. P. Corcoran, M. C. Ionita, G. Costache, 2006Pose invariant face recognition using AAM models. Proceedings of 10 10th International Conference on Optimization of Electrical and Electronic Equipment (OPTIM'06), 4:181-184,.

9. G. N. Costache, 2007 Advances in automated image categorization: Sorting images using person recognition techniques. PhD thesis, Dept of Electronic Engineering, NUI Galway,

10. G.N. Costache, P.M. Corcoran & P. Puslecki, 2009 Combining PCA-based datasets without retraining of the basis vector set, Pattern Recognition Letters, Volume 30, Issue 16, Pages 1441-1447, December,

11. B. Du, S. Shan, L. Qing, W. Gao, 2005Empirical comparisons of sev- eral preprocessing methods for illumination insensitive face recognition. Acoustics, Speech, and Signal Processing, 2 981 984

12. M. Gokmen, E. Vucini, M. E. Groller, 2007 Face recognition under varying illumination. 15 WSCG, 57 64 ,

13. R.C. Gonzalez and R.E. Woods. 1992Digital image processing. Addison-Wesley, Reading, MA,

14. R. Hummel, 1977 Image enhancement by histogram transformation. Computer Graphics and Image Processing, 2 6 184 195

15. M. C. Ionita, 2008Advances in the design of statistical face modelling techniques for face recognition. PhD thesis, Dept of Electronic Engineering, NUI Galway,

16. B.H. Juang and L.R. Rabiner. 2005Automatic speech recognition- a brief history of the technology development. Elsevier Encyclopedia of Language and Linguistics, Second Edition,

17. T. Kanade, 1977 Computer recognition of human faces. Interdisciplinary Systems Research, 47,

18. S. Agazzi, 1994 Keyword spotting in poorly printed documents using pseudo 2-d hidden markov models. IEEE Transactions on Pattern Analysis and Machine Intelligence, 16 842 848 , August 1994.

19. H. S. Le , H. Li, 2004 Face identification system using single hidden markov model and single sample image per person. IEEE International Joint Conference on Neural Networks, 2004.

20. T. M. Lehmann, C. Gnner, K. Spitzer, 1999 Survey Interpolation methods in medical image processing IEEE Transactions on Medical Imaging, 18 1049 1075

21. V. Nefian, 1999 A hidden markov model based approach for face detection recognition PhD Thesis,

22. A.V. Nefian and M.H. Hayes III. 1998 Face detection and recognition using hidden markov models. Image Processing, ICIP 98, Proceedings. International Conference on,, 1 141 145 , October 1998.

23. A.V. Nefian, and M.H. Hayes III 1999, Face recognition using an embedded HMM, IEEE Conf. on A/V-based Biometric Person Authentication, pp. 19-24,.

24. A.V. Nefian and M.H. Hayes III. 2000 Maximum likelihood training of the em- bedded hmm for face detection and recognition. International Conference on Image Processing, 1 33 36 .

25. OpenCV 2006Online Reference Manual, Intel Corporation

26. J. O'Toole, P. J. Phillips, F. Jiang, J. Ayyad, N. Penardand, H. Abdi, 2007 Face recognition algorithms surpass humans matching faces over changes in illumination. PAMI(29), 29(9):1642 EOF 6 EOF 42-1646,

27. P. J. Phillips, M. Hyeonjoon, S. A. Rizvi, P. J. Rauss, 2000 The feret evaluation methodology for face-recognition algorithms. IEEE Transactions on Pattern Analysis and Machine Intelligence, 10 22 1090 1104

28. W. K. Pratt, 1991Digital Image Processing New York, John.Wiley Sons

29. L. R. Rabiner, 1989A tutorial on hidden markov models selected applications in speech recognition Proc. of IEEE, 77 2 257 286 , February

30. L.R. Rabiner and B.H. Juang. 1986 An introduction to hidden markov models. IEEE ASSP Mag, 1 3 4 16

31. E. Rentzeperis, A. Stergiou, A. Pnevmatikakis, L. Polymenakos, 2006 Impact of face registration errors on recognition. Artificial Intelligence Applications and Innovations, 204 187 194

32. F. Samaria, 1994 Face recognition with hidden markov models. Ph.D. thesis, Department of Engineering, Cambridge University, UK,

33. F. Samaria, F. Fallside, 1993 Face identification and feature extraction using hidden markov models. Image Processing: Theory and Applications, Elsevier, 295 298 .

34. F. Samaria, 1994 AC Harter Parameterization of a stochastic model for human face identification. Applications of Computer Vision, Pro- ceedings of the Second IEEE Workshop on, 77 138 142 , December 1994.

35. W.A.P. Smith and E.R. Hancock. 2005 Single image estimation of facial albedo maps. Lecture Notes in Computer Science, 3704 517 526

36. E. Steinberg, P. Corcoran, Y. Prilutsky, P. Bigioi, M. Ciuc, S. Ciurel, & C. Vertran, 2009 Classification system for digital images using workflow, face detection, normalization, and face recognition, US Patent 7,555,148, June.

37. P. Viola, M. Jones, 2001 Robust real-time object detection presented at 2nd International workshop on Statistical and Computational Theories of Vision, Vancouver, Canada, July 13th,.

38. M. H. Yang, D. Kriegman, N. Ahuja, 2002 Detecting faces in images: A survey. IEEE Transactions on Pattern Analysis and Machine Intelligence, 1 24 34 58

39. W. Zhao, R. Chellappa, A. Rosenfeld, P. J. Phillips, Face recognition A literature survey. ACM Computing Surveys 399 458 .

40. K. Zuiderveld, 1994 Contrast limited adaptive histogram equalization. Graphics gems IV, Academic Press Professional, Inc., 474 485 .

Chapter 5

Complementary feature sets for optimal face recognition

Chandan Singh1, Neerja Mittal2* and Ekta Walia3

1Department of Computer Science, Punjabi University, Patiala 147002, India.

2Central Scientific Instruments Organisation, Sector 30-C, Chandigarh 160030,

India.

3Department of Computer Science, South Asian University, Akbar Bhawan, Chanakyapuri, Delhi 110021, India.

ABSTRACT

In face recognition tasks, one kind of feature set is not adequate to generate superior results; thus, selection and combination of complementary features are crucial steps. In this paper, the fusion of two useful descriptors, i.e., the Zernike moments (ZMs) and the local binary pattern (LBP)/local ternary pattern (LTP), has been proposed. The ZM descriptor consists of good global image representation capabilities besides being invariant to image rotation and noise, while the LBP/LTP descriptors capture the innate details within some local parts of face image and are insensitive to illumination variations. The fusion of these two is observed to incorporate the traits of both of these individual descriptors. Subsequently, in this work, the performance of diverse feature sets of ZMs (i.e., magnitude features, magnitude plus phase features, and the real plus imaginary component features) combined with the LBP/LTP descriptor is analyzed on FERET, Yale, and ORL face databases. The recognition results achieved by the proposed method are approximately 10 to 30% higher than those obtained with these descriptors separately. Recognition rates of the proposed method are also found to be significantly better (i.e., by 8 to 24%) in case of single example image per person in the training.

KEYWORDS

Face recognition, Zernike moments (ZMs), Local binary pattern (LBP), Local ternary pattern (LTP), Invariant image features

1 INTRODUCTION

In recent times, face recognition has become one of the widely used biometric techniques having a number of real-world applications like human-computer interaction, surveillance, authentication, computer vision applications, computer user interfaces, etc. An automatic face recognition system consists of some methods to ascertain a person's identity on the basis of his/her physiological characteristics. The sensitivity of available classifiers to different kinds of disparities such as illumination variation, facial expression, facial occlusion, pose variation, aging, etc. is among the most challenging problems that the researchers face [1].

In order to improve the existing face recognition techniques, discriminative competence of the invariant features selected to represent the face images should be high because, thereafter, classification is performed on the basis of these invariant features only. In literature, the approaches used to represent the face images are classified broadly into two categories, namely, the global feature extraction approaches and the local feature extraction approaches [2]. The global feature extraction approaches are based on the statistical methods, wherein features are extracted from the whole face image. In this category, the subspace-based methods, namely, principal component analysis (PCA), Fisher linear discriminant (FLD), two-dimensional PCA (2DPCA), and two-directional two-dimensional PCA ($2D^2PCA$) [3–7], are some of the popular and most frequently employed techniques. Moment invariants, such as Hu's seven moment invariants and orthogonal rotation invariant moments such as Zernike moments (ZMs), pseudo-Zernike moments (PZMs), and orthogonal Fourier-Mellin moments (OFMMs), are observed to be very effective in global image description and recognition [8], and MPEG-7 uses some of them as region-based shape descriptors for image retrieval [9]. The magnitude of these moments is invariant to image rotation, and after applying some geometric transformations, it becomes invariant to translation and scale [10, 11].

Local feature extraction approaches deal with fine information within the specific parts of face images such as eyes, nose, mouth, etc. Recently, a lot of work has been done on these methods because the local features are known to be robust against illumination, occlusion, expression, and noise variations. The local feature extraction approaches have been classified into two categories, i.e., the sparse descriptors and the dense descriptors. The sparse descriptors initially divide a face image into patches and then determine its invariant features. A prominent descriptor in this category is the scale-invariant feature transform (SIFT) introduced by Lowe [12], which possesses useful characteristics of being invariant to scale and rotation. Soyel et al. used the discriminative SIFT (D-SIFT) approach for optimal facial expression recognition, but this method is somewhat susceptible to the illumination variation [13]. In face recognition technology, Gabor wavelet is one of the most frequently used and successful local image descriptors. It incorporates the characteristics of space and frequency domains. The local features extracted by Gabor filters are invariant to scale and orientation and are able to detect edges and lines in the face images [14]. The main difficulty with Gabor filters is their high computational complexity. In case of the dense descriptors, local binary pattern (LBP) is one of the most widely used approaches due to its invariance to monotonic gray-level changes and ease in extraction of the local features. Apart from texture analysis, it has provided excellent results in many areas of image processing and computer vision including its wide use in face recognition [15, 16]. Several variants of LBP are available in literature to represent face images with compact feature sets. Such variants also improve the classification performance of the basic LBP approach [17, 18].

In complex applications like face recognition, it is observed that one kind of feature set is not rich enough to capture the entire face information. Thus, finding and combining the complementary feature sets have become an active research topic in recent years. Specifically, global features are related to the holistic characteristics of face, whereas local features describe the finer details within face images, so it seems logical to combine both of these feature sets since the information conveyed by them belongs to different attributes of the face images. In recent times, many researchers are developing the face recognition algorithms by combining the multiple feature sets. Kim et al. [19] have proposed a combined subspace-based approach using both global and local features obtained by applying linear discriminant analysis (LDA)-based method for face recognition. Zhou and Yang [20] have proposed fusing feature Fisher classifier (F^3C) approach where the face image is first divided into smaller subimages and then the discrete cosine transform (DCT) technique is applied to the whole image and some subimages to extract the holistic and local facial features. After concatenating these DCT-based holistic and local facial features, the enhanced Fisher linear discriminant model (EFM) has been employed to generate a low-dimensional feature vector. Similarly, local and global information extracted by using DCT coefficients along with the Fisher classifier developed for high-dimensional multiclass problem have been proposed in [21]. Singh et al. proposed a robust two-stage face recognition approach by the fusion of global ZMs and Weber law descriptor (WLD)-based local features [22]. The usefulness of combining the global and local facial features is presented in [23] where a hierarchical ensemble of global and local features is performed. In this technique, 2D Fourier transform is used to extract the global features and the Gabor wavelet is opted to extract local features. Subsequently, equal weights are assigned to both the global and local features for combining the outputs of two classifiers (although it is established by the authors that the contribution of both global and local features is different). Wong et al. have proposed dual optimal multiband feature (DOMF) method for face recognition in which wavelet packet transform (WPT) decomposes the image into frequency subbands and the multiband feature fusion technique is incorporated to select optimal multiband feature sets that are invariant to illumination and facial expression. In this method, parallel radial basis function (RBF) neural networks are used to classify the two sets of features. The decision scores are then combined and processed by an adaptive fusion mechanism [24]. The use of steerable pyramid decomposition (S-P transform) both in global and local appearance and feature/score fusion has been analyzed in [25]. In this work, each face image is described by a subset of band-filtered images containing steerable pyramid coefficient. These S-P subbands are divided into small subblocks to extract the compact and meaningful feature vectors that provide a better representation of the class information. Recently, Liu and Liu [26] have proposed an approach for face recognition that fuses color and local spatial and global frequency information. This method is composed of multiple features of face images derived from LBP, DCT, hybrid color space, and the Gabor image representation. The combination of Gabor and LBP enhances the power of spatial histogram that is impressively insensitive to appearance variations. This method has proven to be robust against illumination, pose, and expression variations [27, 28]. However, the combination of Gabor and multiresolution LBP descriptors requires significantly greater computation time.

Although a lot of research is going on combining the multiple feature sets, still the selection of complementary feature sets for fusion and the techniques for combining these divergent feature sets are a challenge. In view of that, in this paper, a fusion of two complementary feature sets is proposed, where the global information of the face images is extracted by the ZM descriptor employing its rotation invariance characteristic, while the LBP/LTP descriptor captures the significant local information. Among various global shape descriptors, ZMs are observed to be one of the best shape descriptors because of their many attractive characteristics [8]. They possess minimum information redundancy, rotation invariance of their magnitude, robustness to noise, etc. The estimation of head movement by using the phase coefficients of ZMs of the original image and that of the rotated image has been employed to generate a set of features that is significantly tolerant to pose variation as well [29]. The magnitude features of ZMs obtained at some higher order of moments are observed to be invariant to expression variation [30]. On the other hand, the LBP descriptor is observed to be relatively more insensitive to illumination changes. It is computationally efficient as well as quite simple to implement. Recently, a useful extension to this approach is introduced, namely, the local ternary patterns (LTP), that is observed to be more discriminative and invariant to image noise in near-uniform regions as compared to LBP [18]. Particularly, combining the feature sets that are invariant to global variations as well as to local changes of face images would be an effective approach to achieve an optimal face recognition system. As discussed earlier, the information conveyed by the ZM and LBP/LTP descriptors are distinct and belong to different aspects of a facial image. Fusion of these descriptors is expected to be enriched with the useful characteristics of both of them. One of the critical issues involved in the fusion process is the time spent in the computation of the combined features. It is shown through time analysis of the proposed approach that the total time required for the recognition process is very small and can be afforded by PCs and other low computation devices.

The ZM descriptor provides three different sets of features, namely, magnitude features, combined magnitude and phase features [31, 32], and the modified real and imaginary component features [29]. In this study, these diverse feature sets of ZMs are referred to as ZM_{mag}, $ZM_{magPhase}$, and $ZM_{component}$, respectively. The performance of the feature sets of ZMs combined with LBP descriptor, in comparison to that of the ZMs coupled with the LTP descriptor, is also analyzed. Consequently, the proposed fusion of the diverse feature sets of ZMs and the LBP/LTP descriptors provides various combined approaches such as $ZM_{mag}LBP$, $ZM_{magPhase}LBP$, $ZM_{component}LBP$, $ZM_{mag}LTP$, $ZM_{magPhase}LTP$, and $ZM_{component}LTP$. In order to compare the performance of these combined approaches to that of the individual ZM and LBP/LTP approaches, exhaustive experiments are performed on three prominent face databases, namely, FERET, Yale, and ORL, against pose, illumination, expression, and noise variations. The results obtained show that the recognition rate of the combined approaches, in comparison to that of their individual counterparts, is significantly better varying between 10 and 30%. Experimental results also prove the efficacy of the proposed methods over other existing works. A significant improvement in recognition rate is achieved for the case of single training image per person.

The rest of the paper is organized as follows: Section 2 presents a brief overview of the ZM approach and the diverse feature sets obtained from it and includes a brief introduction to the LBP/LTP approaches, Section 3 describes the similarity measures used to evaluate the matching score of these methods, the procedure involved in the proposed fusion of the ZM and the LBP/LTP descriptors is described in Section 4, the experiments and results obtained are presented in Section 5, and the conclusions and future directions are presented in Section 6.

2 BASELINE IMAGE DESCRIPTORS

2.1 Global image descriptor

2.1.1 Zernike moments

The Zernike functions constitute a set of orthogonal basis functions mapped over the unit circle. Zernike moments of a function *f* (*x*, *y*) are constructed by projecting it onto those functions. The ZMs of order *n* and repetition *m* are defined by

$$Z_{nm} = \frac{n+1}{\pi} \iint_{x^2+y^2\leq 1} f(x,y) V^*_{nm}(x,y) dxdy \tag{1}$$

where $n \geq 0$, $|m| \leq n$, and V∗nm(x,y)$V^*_{nm}(x,y)$ are the complex conjugates of the Zernike function $V_{nm}(x, y)$, where

$$V_{nm}(x,y) = R_{nm}(x,y) e^{jm\theta} \tag{2}$$

with $= \sqrt{-1}$, $\theta = \tan^{-1}(y/x)$, $\theta \in [0, 2\pi]$, and

$$R_{nm}(x,y) = \sum_{s=0}^{(n-|m|)/2} \frac{(-1)^s (n-s)! (x^2+y^2)^{\frac{n-2s}{2}}}{s! \left(\frac{(n+|m|)}{2} - s\right)! \left(\frac{(n-|m|)}{2} - s\right)!} \tag{3}$$

The ZMs are derived for a discrete image function using zeroth-order approximation of Equation 1 given by [10]

$$Z_{nm} = \frac{n+1}{\pi} \sum_{\substack{i=0 \\ x_i^2+y_k^2\leq 1}}^{N-1} \sum_{k=0}^{N-1} f(i,k) V^*_{nm}(x_i, y_k) \Delta^2 \tag{4}$$

where

$$x_i = \frac{2i+1-N}{N\sqrt{2}}, y_k = \frac{2k+1-N}{N\sqrt{2}},$$
$$i,k = 0,1,2,\ldots,N-1, \text{ and } \Delta = \frac{2}{N\sqrt{2}} \tag{5}$$

2.1.2 Diverse feature sets of ZMs and related work

Since the magnitude of ZMs is invariant to rotation, usually it is used as invariant image descriptor in many image analysis and pattern recognition applications. The phase component of ZMs is, however, ignored. It is observed that the phase component also carries equally significant information as the magnitude component does [31]. Therefore, in recent years, significant research work has been carried out to incorporate ZM phase coefficients along with their magnitudes as invariant feature descriptors. At present, there are two approaches to realize this objective. In the first approach, developed by Revaud et al. [33], a similarity measure incorporating both the magnitude and phase coefficients of the query and database image is used. The method provides excellent pattern matching performance but at the cost of enhanced computation time. In the second approach, the rotation angle between a query image and the database image is estimated. It is assumed that the query image is the rotated version of the original database image. The estimated rotation angle is used to cancel the effect of rotation in order to compare the phases of the query and database images. Recently, we devised a novel way to correct phase coefficients without estimating rotation angle. The method was applied successfully in face recognition [29]. The method works as follows: Suppose Z_{nm} and Z'_{nm} are the ZMs of database and query images, respectively, and ϕ_{nm} and ϕ'_{nm} are their respective phase angles. We compute $m\theta = \phi_{nm} - \phi'_{nm}$ and correct the ZMs of the query image by evaluating $Z'^{c}_{nm} = Z'_{nm} e^{jm\theta}$. If the two images are same, then $Z'^{c}_{nm} = Z_{nm}$, otherwise $Z'^{c}_{nm} \neq Z_{nm}$; therefore, the real and imaginary components of ZMs of the query and database images can be compared separately, instead of comparing only their magnitude. An attractive advantage of this approach is that by using two-component feature vectors, the number of features is almost doubled as compared to the ZM magnitude only features for the same order of moments. This approach has additional advantages of having low computation cost, less susceptibility to image noise, and numerical stability, in addition to providing better recognition rate [29]. Throughout the paper, these features of ZMs based on magnitude, magnitude together with corrected phase [31], and the corrected real and imaginary parts of ZMs [29] are referred to as ZM_{mag}, $ZM_{magPhase}$, and $ZM_{component}$, respectively.

2.2 Local image descriptor

2.2.1 Local binary pattern

Ojala et al. have introduced the local binary patterns (LBP) for effective texture description that has been used in many image processing and computer vision applications [15]. The most important property of this approach is its tolerance against illumination variation. Being computationally simple, it provides significant advantage over other approaches. The LBP operator takes some specific neighborhood around each pixel, and it then thresholds the values of these neighborhood pixels with respect to the central pixel's value. The resulting binary pattern is used as an element of the local image descriptor. Thus, it assigns a label to every pixel p_i of an image by thresholding its respective 3 × 3 neighborhood values with the value of the central pixel p_c and producing the result in the form of an 8-bit binary code. The LBP operator is computed as

$$\text{LBP} = \sum_{i=0}^{7} 2^i b(p_i - p_c) \tag{6}$$

$$b(p_i - p_c) = \begin{cases} 1, & \text{if } p_i \geq p_c \\ 0, & \text{otherwise} \end{cases} \tag{7}$$

where the values of i move along the eight neighbors of the central pixel. In case of 8-bit patterns, Ojala et al. [15] have observed that out of 2^8 patterns, only 58 uniform patterns provide approximately 90% information of the image neighborhoods while the remaining patterns consist of mostly noise. This attribute significantly reduces the number of LBP histogram bins from 256 to 59 where all the non-uniform patterns are stored in a single bin, the 59th bin.

2.2.2 Local ternary patterns

The local histogram features obtained from LBP have proven to be highly discriminative in face recognition [16, 17]. However, they are found to be sensitive to noise because of the fact that they are thresholded exactly at the value of the central pixel especially in near-uniform and smooth regions of face images like cheeks or forehead. Recently, an important extension to the original LBP is provided by Tan et al. [18]. It generates three-valued codes corresponding to each image pixel. In their method, the binary LBP code is replaced with the ternary LTP code and the gray values in a zone of width $\pm w$ around the central pixel p_c are quantized to 0 and the values above this zone are quantized to +1 while those below it are quantized to −1. Specifically, the value of $b(x)$ given in Equation 7 is replaced with the following three-valued function:

$$b'(p_i, p_c, w) = \begin{cases} 1, & p_i \geq (p_c + w) \\ 0, & |p_i - p_c| < w \\ -1, & p_i \leq (p_c - w) \end{cases} \tag{8}$$

where w is a user-defined threshold. The LTP code is assumed to be invariant to image noise but may not be strictly invariant to gray-level transformations. The concept of uniform patterns to obtain histogram features is also applicable to LTP. For simplicity, the three-valued LTP codes are split into their positive and negative bisects which generate two sets of histogram features out of which one corresponds to the positive patterns and the other represents the negative patterns [18].

3 SIMILARITY MEASURES USED

In this section, the similarity measures used for finding the matching scores of ZM and LBP/LTP descriptors are discussed briefly. In this work, it is observed that the fusion of matching scores obtained by applying L_2 – Norm/L_1 – Norm on the ZM descriptor and histogram intersection on the LBP/LTP descriptor generates superior performance. Hence, these different similarity measurement techniques are used on the feature sets generated by these descriptors. Since the matching scores obtained from these different approaches are heterogeneous, normalization is required to transform these matching scores to a common range before combining them.

3.1 Similarity measure for ZM descriptor

The magnitude features of ZMs, i.e., ZM_{mag}, of two images are compared by evaluating the normalized Euclidean distance (L_2 – Norm) between them. The normalized L_2 – Norm between the two sets of feature vectors of ZMs is given by

$$d_{\text{mag}} = \frac{1}{L}\sqrt{\frac{\left(|Z_i'|-|Z_i|\right)^2}{\max\left(|Z_i'|^2, |Z_i|^2\right)}} \tag{9}$$

where Z'i and Z_i are the feature vectors of the query and the database images, respectively, and L represents the size of the feature vector consisting of the magnitude of ZMs. The normalized Euclidean distance d_{Phase} defined by [31], between ZM phases of the query and the database images, is computed as

$$d_{\text{Phase}} = \frac{1}{L}\sqrt{\sum_{i=1}^{L}\frac{1}{2\pi}\left(|\varphi_i'|-|\varphi_i|\right)^2} \tag{10}$$

where φ is the phase angle of the database image and φ' is the phase angle after estimating the rotation angle between the query and database images and correcting the phase [31]. The total distance d_{magPhase} between the feature vectors consisting of ZM_{magPhase} coefficients has been evaluated by using the distances d_{mag} and d_{Phase}, computed as per Equations 9 and 10, respectively. The formula used to compute the d_{magPhase} is given as

$$d_{\text{magPhase}} = \left(w_1 d_{\text{mag}} + w_2 d_{\text{Phase}}\right)/\left(w_1 + w_2\right) \tag{11}$$

Normally, equal weights are assigned to simplify this process, i.e., $w_1 = w_2 = 0.5$.

The $ZM_{\text{component}}$ descriptor includes the modified real and imaginary parts of ZMs to formulate a two-component feature vector for each ZM. The normalized L_1 – Norm-based distance measure for the evaluation of similarity between component features of the database and the query images is given as under

$$d_{\text{comp}} = \frac{1}{L}\sum_{i=1}^{L}\frac{\left(|Re(Z_i)-Re(Z'^{c}_{i})|\right)}{\max\left(|Re(Z_i)|, |Re(Z'^{c}_{i})|\right)} + \frac{\left(|Im(Z_i)-Im(Z'^{c}_{i})|\right)}{\max\left(|Im(Z_i)|, |Im(Z'^{c}_{i})|\right)} \tag{12}$$

The above mentioned distance metric d_{comp} has proven to be a better similarity measure between two sets of component feature vectors [29].

3.2 Similarity measure for LBP/LTP descriptor

The histogram intersection distance has been used to compare the feature vectors of query and database images for both the LBP and the LTP descriptors. The histogram intersection distance evaluated for every bin *n* between the database image and the query image is given as

$$D_h(Hd, Hq) = \frac{\sum_{n=1}^{B} \min(Hd_n, Hq_n)}{\min\left(\sum_{n=1}^{B} Hd_n, \sum_{n=1}^{B} Hq_n\right)}, \quad D_h(Hd, Hq) \in [0, 1] \tag{13}$$

where *Hd* and *Hq* are the histograms consisting of LBP/LTP features of the database and the query images, respectively. *B* is the total number of bins in the histograms. If either of the two images, i.e., the database and the query images is identical, then the value of $D_h(Hd, Hq)$ is 1.

4 FUSION OF ZM AND LBP/LTP DESCRIPTORS

The ZM descriptor and the LBP/LTP descriptors are observed to be complementary to each other, and their fusion is expected to be able to discriminate the face images even in the presence of diverse variations. The ZM descriptor is observed to extract the global information of the images more effectively as compared to that of any other global descriptor [9]. On the other hand, the LBP and the LTP descriptors have been established to be successful methods for representing the finer interior details within the face images. The feature set established by the fusion of these autonomous approaches, i.e., ZMs and the LBP/LTP, is supposed to be enriched with the invariant characteristics of both of them. Exhaustive experiments performed against pose, illumination, expression, and noise variations on the suitable databases prove that the said hypothesis is correct.

The procedure followed to recognize the face images by the proposed combined approaches, i.e., ZM_{mag}LBP, $ZM_{magPhase}$LBP, $ZM_{component}$LBP, ZM_{mag}LTP, $ZM_{magPhase}$LTP, and $ZM_{component}$LTP, is briefly described in Figure 1. The recognition of face images through the proposed fusion of feature sets includes three stages - feature extraction, fusion of similarity score, and classification. The first stage of this procedure creates the invariant feature sets extracted by using ZM and LBP/LTP descriptors. The second stage involves fusion of the matching scores obtained from these feature sets after applying the similarity measures as described in the previous section. A number of feasible techniques such as fusion at the feature extraction level, matching score level, or decision level exist for combining the multiple feature sets. It is not easy to combine the information at the feature level when the feature sets obtained by different techniques are either inaccessible or incompatible. Fusion at the decision level is too rigid as only a limited amount of information is available at this level. Therefore, integration at the matching score level is generally preferred due to the ease of accessing and combining matching scores [34]. In the proposed work, feature vectors are obtained by applying ZM and LBP/LTP descriptors which provide complementary information. Further, we observed that for the LBP/LTP descriptor, the matching score evaluated by the histogram intersection measure gives better results than using L_2 – Norm. Hence, in this work, the fusion at the matching score level is employed wherein the histogram intersection (using Equation 13) and L_2 – Norm (using Equations 9 and 11 for ZM_{mag} and $ZM_{magPhase}$)/L_1 – Norm (using Equation 12 for $ZM_{component}$) is used for evaluating the matching scores of the feature vectors obtained from the LBP/LTP and ZM descriptors, respectively. Thereafter, these individual matching scores are combined by using the sum rule to generate a single scalar score which is then used to make the final decision. The sum rule to compute the fusion of individual matching scores is given as below

$$\text{sum rule } (F_{\text{sr}}) = \frac{\{S_{\text{Z}} + (1 - S_{\text{L}})\}}{2} \tag{14}$$

where S_Z and S_L represent the matching scores of the ZM descriptor and the LBP/LTP descriptors, respectively. The matching scores of these approaches are normalized before fusion. Normalization is required to map the matching scores obtained from multiple frameworks to a common range so that they can be easily combined. In order to combine these matching scores, S_L is subtracted from 1, so that the histogram intersection would now signify higher similarity with lower values. Finally, in the third stage, we use the nearest neighbor rule to perform classification. This method always gives us only one recognized image which is labeled as either correct or incorrect in order to evaluate the recognition performance. The recognition rate (in percentage) is measured by using the following formula:

$$\text{RecRate} = \frac{(N\text{test} - Nf)}{N\text{test}} \times 100 \tag{15}$$

where N test is the total number of images in the test set and Nf is the number of images recognized incorrectly.

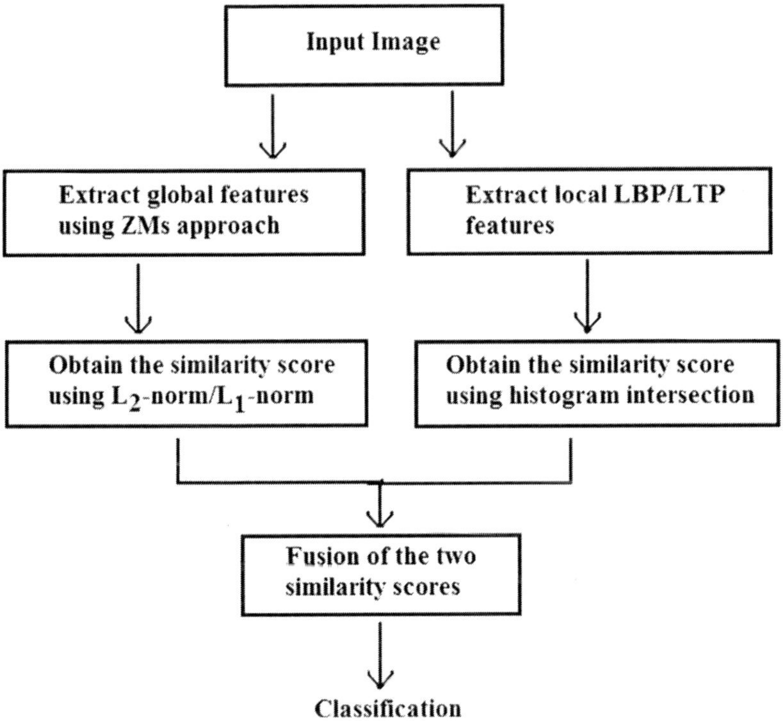

Figure 1. Flow chart depicting the procedure involved in fusion of the ZM and LBP/LTP descriptors.

5 EXPERIMENTS AND RESULTS

In order to evaluate the performance of the considered autonomous approaches in comparison to that of the proposed combined approaches, experiments are performed on three well-known and calibrated face databases, namely, FERET face database [35] consisting of images in diverse variations, Yale face database [36] consisting of illumination and expression variations, and ORL face database [37] having small pose (tilt/yaw) changes. It is well known that the accuracy of the face recognition system is significantly affected by the kind of variations present in images of the face database as well as by the number of images of each subject (i.e., person) in the training set. Thus, exhaustive experiments in a comprehensive and deterministic manner are performed with respect to different types of variations present in these databases. The number of training images per person is also varied to observe its effect on recognition accuracy. The best results are highlighted in italics. All the experiments are performed in Visual C++6.0 under Microsoft Windows environment on a PC with a 3.0-GHz CPU and a 3-GB RAM.

5.1 Performance on FERET database

FERET grayscale face database has become the most popular and standard database in the field of face recognition. We have performed experiments on two subsets of this database, consisting of the frontal to profile pose variation. The first subset is formed by randomly selecting 100 persons with seven different poses (yaw) 0°, ±22.5°, ±67.5°, and ±90°. The second subset consisted of FERET *'b'* category images of 200 persons in different illumination, expression, and pose angles of 0°, ±15°, ±25°, ±40°, and ±60°. In this work, the first subset is called FERET_A. It consists of 700 images. The second subset is named as FERET_B and contains 2,200 images. FERET evaluation protocol partitions the database into gallery (1,196 images of 1,196 persons) and four probe sets, namely, *fafb*, *fc*, *dup I*, and *dup II*. The images in the *fafb* set are with facial expression variation, the *fc* set contains images with illumination variations, and the images in the *dupI* and *dupII* sets represent aging effects. For detailed experimentation against pose, expression, and illumination variations, various data partitions are generated for these subsets which are described in Table 1. The original images of this database are of size 256 × 384 pixels. We transformed these images into 128 × 128-pixel size in order to reduce the time taken for conducting the experiments, while the face images from FERET_Gallery/Probe subset are cropped and resized to 64 × 64 pixels. Some sample images, for one person, from this database are shown in Figure 2. The face images of FERET_A, FERET_B, and FERET_Gallery/Probe subsets are partitioned into 64 patches of 16 × 16 and 8 × 8 pixels, respectively, to extract the local LBP/LTP features, while the global ZM features are extracted from whole face images.

In order to analyze the performance of the proposed combined approaches, the first set of comprehensive experiments is performed on FERET_A1 category as described in Table 1. The different possible trials for this setup, containing various combinations of the training and the test sets, are shown in Table 2. The average recognition performance of the individual and the combined approaches over these different trials (seven) is presented in Figure 3 for different values of maximum order of ZMs, denoted by $n_{\max}$. Further, it is pertinent to mention here that the values of $n_{\max}$ have no effect on the performance of the LBP and LTP descriptors. So, the results presented for the LBP/LTP descriptors remain the same for each value of $n_{\max}$. From the results presented, it is observed that among the autonomous approaches, the performance of the $ZM_{component}$ approach is better than that of others.

However, the proposed combined approaches exhibit significantly high recognition rates compared to their individual counterparts. In this experiment, the $ZM_{magPhase}LBP$ approach provides the highest recognition rate at 71.24%.

Next, experiments are performed on FERET_A2 category, and the recognition results are presented in Figure 4 for both the individual and the combined approaches over different order of moments n_{max} used for the ZM descriptors. The basic LBP/LTP descriptors are not invariant to rotation; however, in this category, they perform better than the ZM descriptor. This is due to the fact that the higher pose angles occlude a significant portion of the face and on this kind of distortion, the local feature sets are observed to be more successful than the global features. On the other hand, the proposed fusion of the ZM descriptors and the LBP/LTP descriptors achieves approximately 20% improvement in the recognition results in comparison to that of these independent approaches. It is also noticed that the ZM descriptors coupled with the LBP descriptor generate superior results and the highest recognition rate of 77.33% is accomplished by the $ZM_{magPhase}LBP$ approach. In case of the LTP descriptor, the LTP combined with the $ZM_{component}$ descriptor, i.e., $ZM_{component}LTP$, provides better results than those of other combinations.

Further experiments are performed on FERET_A3, FERET_A4, and FERET_A5 categories. The recognition results for these setups are shown in Table 3. On this database, fusion of the ZM features obtained for $n_{max} = 9$ provides better results. Accordingly, here, all other experiments have been conducted only for this order of moments. It is observed from Table 3 that there is an improvement in recognition rates by approximately 10 to 20% due to the fusion of the ZM and LBP/LTP descriptors. Further, it is observed that the recognition rates decline significantly with increase in the pose angle of test images (e.g., the highest recognition rate of only 74.5% is noticed for FERET_A4 which contains pose variations of ±67.5° in test images). This outcome is obvious because of the fact that the presence of the higher pose angle will occlude a significant part of the face image. The highest recognition rate of 86.5% is achieved by the $ZM_{magPhase}LBP$ approach on FERET_A3 category. On FERET_A5 category, a superior recognition rate of 99.5% is achieved by both the $ZM_{component}LBP$ and the $ZM_{component}LTP$ approaches. Similarly, the $ZM_{magPhase}LBP$ approach provides the highest recognition rate on FERET_A4 category which reveals the contribution of ZM phase coefficients towards the improvement in recognition results.

Table 1. Data partition on FERET database for performing various experiments

Set of experiment	Category	Training	Testing	Remarks
1	FERET_A1	One image of each person resulting in a total of 100 images	Remaining six images of each person in different pose variations, i.e., a total of 600 images	Seven different trials of this setup (as shown in Table 2) have been taken, and the recognition result given in Figure 3 is the average of all these trials
2	FERET_A2	One image of each person in frontal, i.e., 0° pose	Six remaining images having pose variations of ±22.5°, ±67.5°, and ±90°	Against the frontal pose image, all the images in different poses (≤ ± 90°) are taken for testing
3	FERET_A3	One image in frontal pose	Four images having pose variations of ±22.5° and ±67.5°	Testing the performance for small and large pose variation up to ±67.5°
4	FERET_A4	One image in frontal pose	Two images in ±67.5° pose angle	Testing the performance for large pose variation in the left and right directions
5	FERET_A5	One image in frontal pose	Two images in ±22.5° pose angle	Testing the performance for small pose variation in the left and right directions
6	FERET_B1	One image in frontal pose, i.e., the image labeled *'ba'*, is in the training set	Ten remaining images having illumination, expression, and pose (up to ±60°) variations, i.e., the images labeled *bk*, *bj*, *bb*, *bc*, *bd*, *be*, *bf*, *bg*, *bh*, and *bi*, are kept in the test set	Images consisting of all the three variations, i.e., illumination, expression, and pose (up to ±60°), are taken for testing against the frontal pose image. A total of 200 images are in the training set, and 2,000 images are in the test set
7	FERET_B2	Three random images of each person resulting in a total of 600 images	Remaining seven images of each person in different variations, i.e., 1,600 images	Ten different trials of this setup have been taken, and the recognition result (Table 3) is the average of all these trials

8	FERET_B3	One image of each person in frontal pose, i.e., the image labeled *'ba,'* is in the training set and contains a maximum of 200 face images in it	Eight images of each person having pose variations (up to ±60°), i.e., the images labeled *bb*, *bc*, *bd*, *be*, *bf*, *bg*, *bh*, and *bi*, are kept in the test set resulting in a maximum of 1,600 images in it	This category consists of the images of 200 persons in 9 pose variations, from frontal to profile pose, resulting in a total of 1,800 images in it. In this category, the different experiments are carried out by varying the number of subjects (persons) in the database
9	FERET_Gallery/Probe	Standard gallery set *fa* contains 1,196 images of 1,196 subjects in frontal view	*fafb*, *fc*, *dupI*, and *dupII*	The images in the *fafb* (1,195 images) set are with facial expression variation, the *fc* (194 images) set contains images with illumination variation, and the images in the *dupI* (722 images) and *dupII* (234 images) sets represent aging effects

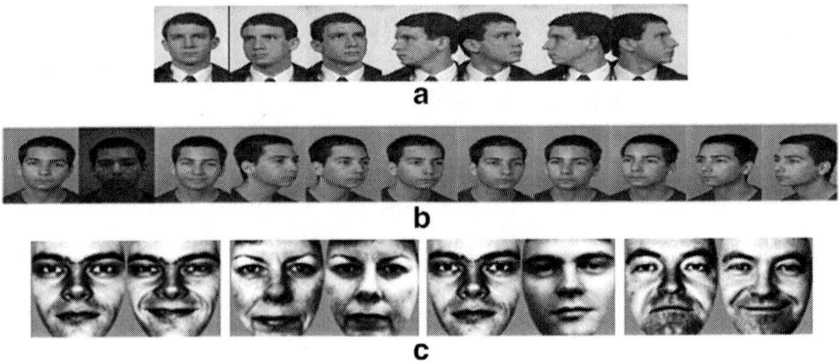

Figure 2. Sample face images. Sample images from **(a)** FERET_A, **(b)** FERET_B, and **(c)** FERET_Gallery/Probe subsets.

Table 2. Different trials for one image in training and remaining six in test set for FERET_A1 category

Trial	Image in training set in pose angle	Images in test set in pose angle
1	0°	±22.5°, ±67.5°, ±90°
2	+22.5°	0°, −22.5°, ±67.5°, ±90°
3	−22.5°	0°, +22.5°, ±67.5°, ±90°
4	+67.5°	0°, ±22.5°, −67.5°, ±90°
5	−67.5°	0°, ±22.5°, +67.5°, ±90°
6	+90°	0°, ±22.5°, ±67.5°, −90°
7	−90°	0°, ±22.5°, ±67.5°, +90°

For FERET_B, the recognition results obtained by performing the experiments on FERET_B1 and FERET_B2 categories are also shown in Table 3. It is clear from the experimental results that the combined approaches exhibit approximately 15 to 20% hike in the recognition rate than that obtained by the individual approaches. The highest recognition rate of 78.5 and 88.2% is achieved by the $ZM_{magPhase}$LBP descriptor for FERET_B1 and FERET_B2 categories, respectively. The result obtained on FERET_Gallery/Probe set ascertains the robustness of the proposed system against changes in expression and lighting; however, further research in aging is required. In general, on FERET_B and FERET_Gallery/Probe subsets, the combination of $ZM_{magPhase}$ with LBP descriptor generates higher results than others.

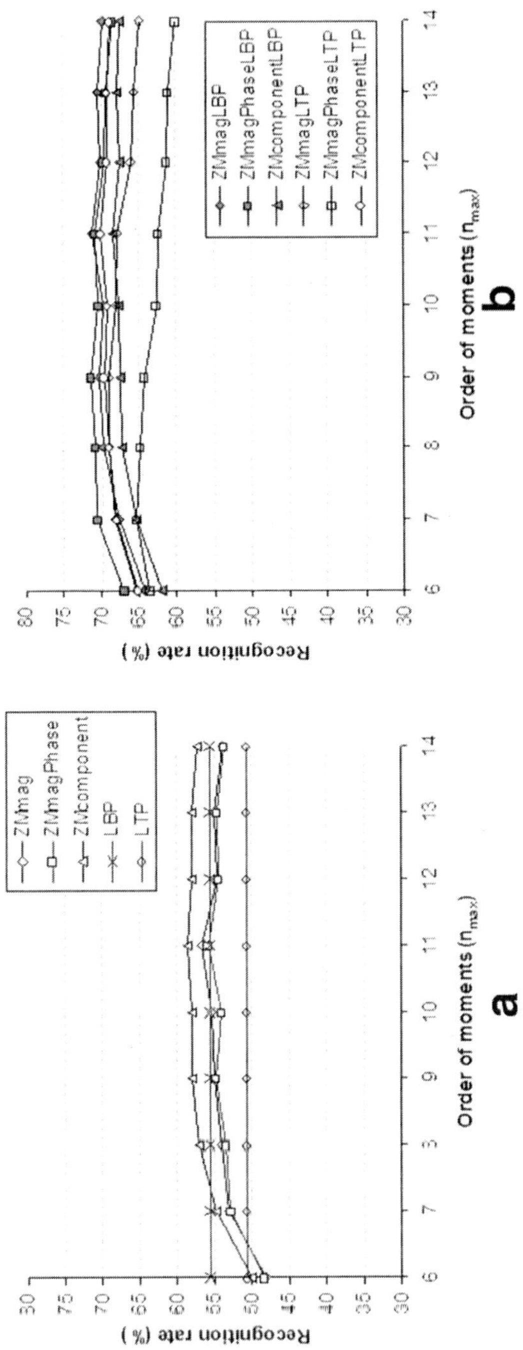

Figure 3. Performance (average) on FERET_A1 category. Average recognition performance of **(a)** individual and**(b)** combined approaches.

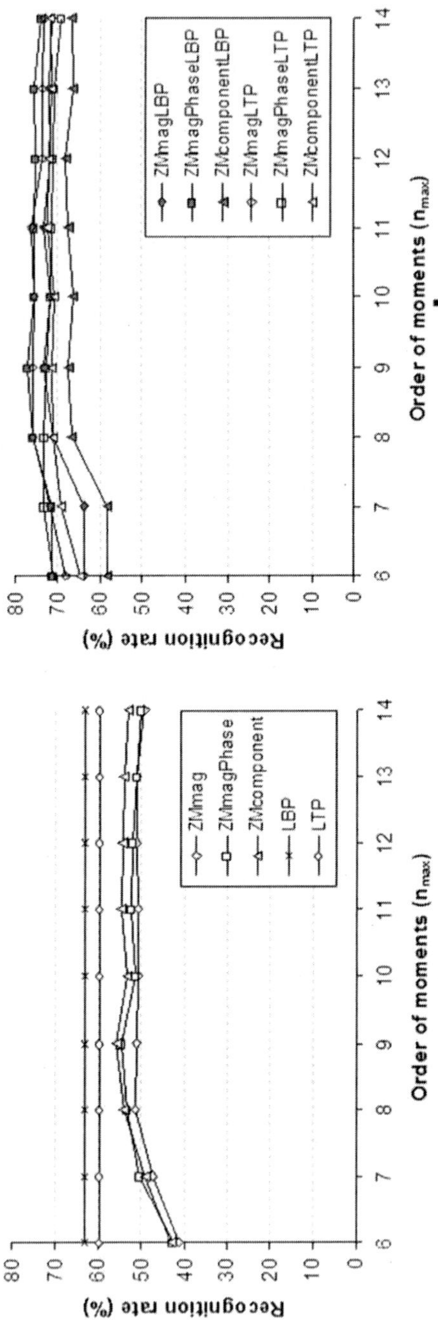

Figure 4. Performance on FERET_A2 category. Recognition results of **(a)** individual and **(b)** combined approaches.

Table 3. Performance of the considered approaches on FERET database

Descriptor	FERET_A3	FERET_A4	FERET_A5	FERET_B1	FERET_B2	FERET_Gallery/Probe			
						fb	*fc*	*dupI*	*dupII*
ZM_{mag}	53.75	34.0	93.5	62.0	76.3	67.53	31.8	35.0	38.0
$ZM_{magPhase}$	67.25	40.0	94.5	60.5	75.87	68.02	32.96	35.8	39.0
$ZM_{component}$	63.75	41.5	96.0	63.45	76.93	65.0	30.36	33.0	37.0
LBP	74.75	56.0	93.5	66.7	80.71	96.87	79.0	66.0	63.7
LTP	70.5	50.0	91.0	64.75	80.08	94.87	72.74	62.0	56.0
ZM_{mag}LBP	83.0	67.0	99.0	77.5	86.43	97.58	89.18	69.25	66.0
$ZM_{magPhase}$LBP	*36.5*	*74.5*	98.5	*78.5*	*88.2*	*98.04*	*91.5*	*69.39*	*66.5*
$ZM_{component}$LBP	79.5	59.5	*99.5*	74.05	84.69	96.0	88.0	68.5	64.0
ZM_{mag}LTP	34.5	70.0	99.0	76.3	85.48	93.0	85.0	65.0	59.5
$ZM_{magPhase}$LTP	81.75	65.5	98.0	75.55	85.97	93.5	85.8	65.7	58.5
$ZM_{component}$LTP	82.25	65.0	*99.5*	74.6	84.74	91.2	84.0	64.0	58.0

5.2 Performance on Yale database

The Yale face database contains 11 images per person for 15 individuals resulting in a total of 165 images. The images in this database have major variations in illumination and facial expressions. They also have images demonstrating occlusion of eyes with eyeglasses. The original size of the images in this database is 243×320 pixels with 256 gray levels. For the experiments, these are cropped down to 64×64 pixels. Sample cropped images from this database, for one person, are shown in Figure 5. Here also, the face images are partitioned into 64 patches of 8×8 pixels to extract the local LBP/LTP features.

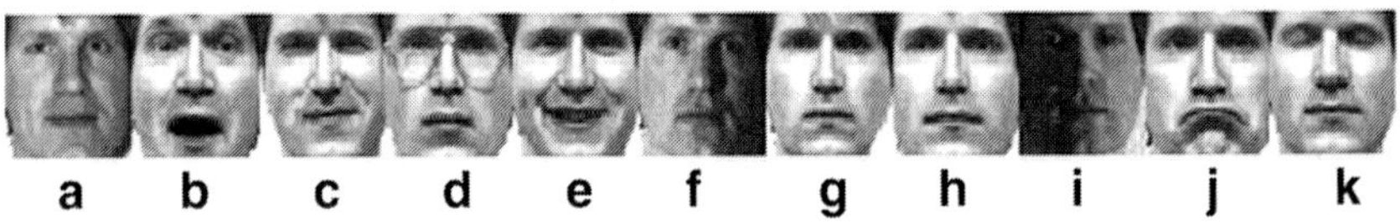

Figure 5. Sample cropped face images (a-k) for one person from Yale database.

In order to examine the improvement in performance by the proposed combined approaches across the expression and illumination variations, exhaustive experiments are performed on this database by taking different number of images in training and test sets. Accordingly, for this purpose, various data partitions have been generated which are presented in Table 4. The first set of comprehensive experiments is performed on YALE 1 category where out of total 11 images of each person, one image is taken in the training set and all the remaining are placed in the test set. This process is repeated 11 times by taking different face images of each person in the training set. Average recognition results over 11 different runs of training and test sets are presented in Figure 6 for different n_{max} of ZMs.

Table 4. Data partition on Yale database for performing various experiments

Set of experiment	Category	Training	Testing	Remarks
1	YALE 1	Random one	All remaining (i.e., remaining ten images except the one selected for the training)	Eleven different trials of this setup have been taken, and the recognition result (Figure 6) is the average of all these trials
2	YALE 2.1	Random two	Remaining nine	The recognition result (Table 5) is the average of ten random trials on each setup, i.e., YALE 2.1, YALE 2.2, YALE 2.3, YALE 2.4, and YALE 2.5
	YALE 2.2	Random three	Remaining eight	
	YALE 2.3	Random four	Remaining seven	
	YALE 2.4	Random five	Remaining six	
	YALE 2.5	Random six	Remaining five	
3	YALE 3.1	a	f, g, h, i	Testing consists of experiments against illumination variation
	YALE 3.2	b, c, d, e, j, k	a, f, g, h, i	
	YALE 4.1	a	b, c, d, e, j, k	Testing consists of experiments against expression variation
	YALE 4.2	f, g, h, i	a, b, c, d, e, j, k	

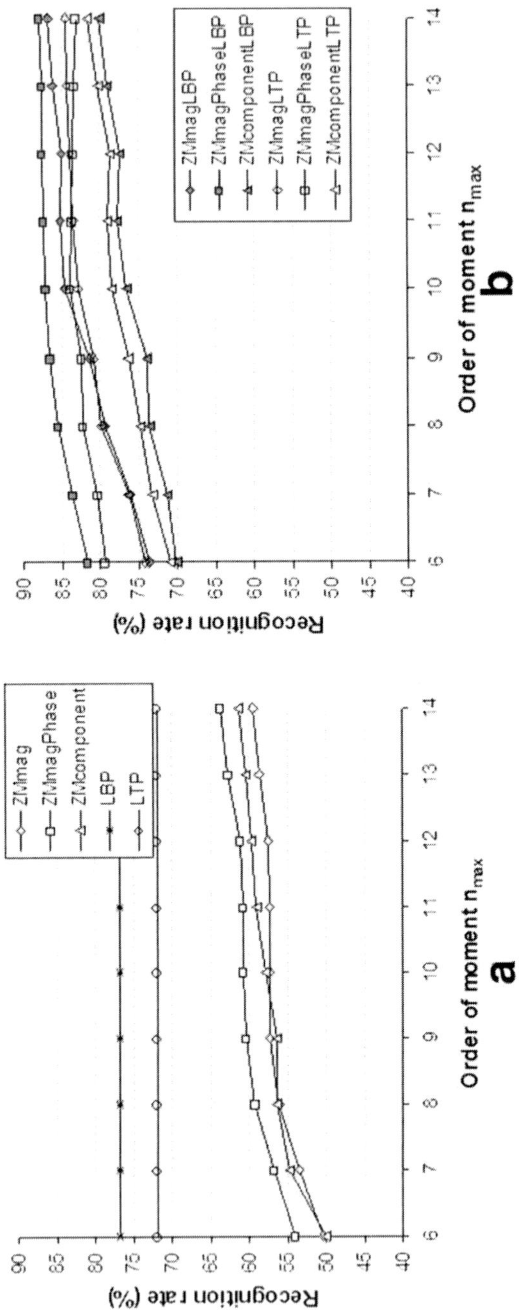

Figure 6. Performance (average) on YALE 1 category against illumination and expression variations. Recognition results of **(a)** individual and **(b)** combined approaches.

From the results obtained, it is observed that an average improvement of approximately 12% is achieved by the proposed combined descriptors as compared to the individual approaches. In case of individual approaches, the performance of both the LBP and LTP approaches is better than that of the three descriptors of ZMs. The result obtained depicts that the local approaches are able to capture the interior details of a face image more efficiently than the global ones. This certainly enhances the suitability of these methods to outperform even in the presence of only a single exemplar image per person. From the results depicted in Figure 6, it is observed that among all the combined approaches, the highest recognition results are achieved by the $ZM_{magPhase}LBP$ descriptor. It is also observed that for the proposed combined methods, fusion of ZM features obtained for $n_{max} = 11$ provides better results. Hence, on this database, all other experiments have been carried out for this order of moments.

Next, the experiments are performed on YALE 2.1, YALE 2.2, YALE 2.3, YALE 2.4, and YALE 2.5 categories. The average recognition results over ten trials of each group of the training and the test sets are presented in Table 5. It is well known that the LBP and LTP descriptors are invariant to changes in intensities of the images, so the results obtained by these two approaches are quite higher than those obtained by the ZM descriptors. Hence, on this database, it has been realized that the LBP/LTP feature sets contribute much more towards the improvement in the recognition rate of the proposed combined approaches which are significantly higher than the individual approaches. In most of the cases, the LBP and LTP descriptors combined with ZM_{mag} features generate superior results. For example, on YALE 2.4 category, the highest recognition rate of 97.56% is achieved by the $ZM_{mag}LBP$ approach. However, on YALE 2.3 category, the average highest recognition rate is 97.14% with the $ZM_{mag}LBP$ approach. Thus, in general, the proposed $ZM_{mag}LBP$ approach outperforms the others.

Thereafter, experiments are performed on YALE 3.1 and YALE 3.2 categories against illumination variation. Similarly, in order to examine the performance of the proposed approaches particularly over expression variation, experiments are carried out on YALE 4.1 and YALE 4.2 categories. The results obtained from these experiments are presented in Table 6. From the results shown, it is clearly noticed that the proposed combined approaches show an improvement in performance by approximately 30% over the ZM descriptors alone, whereas in comparison to the performance of individual LBP/LTP descriptors, an improvement of approximately 10% is achieved. As described earlier, among the individual approaches, the performance of the LBP/LTP descriptors is better than that of the ZM descriptor. Between these two descriptors, the LBP descriptor generates higher recognition rate against the illumination and expression variations on YALE 3.1, YALE 3.2, and YALE 4.2 categories while the

LTP descriptor gives higher results on only YALE 4.1 category for expression variation.

Table 5. Performance (average) of the individual/combined approaches on Yale database

Descriptor	YALE 2.1	YALE 2.2	YALE 2.3	YALE 2.4	YALE 2.5
ZM_{mag}	70.96	74.42	79.43	80.56	84.27
$ZM_{magPhase}$	74.37	79.08	81.62	81.33	84.53
$ZM_{component}$	74.29	77.75	81.05	81.22	83.6
LBP	89.11	89.67	93.14	92.89	93.33
LTP	85.85	89.75	92.57	94.22	93.73
$ZM_{mag}LBP$	93.19	94.83	97.14	97.56	98.27
$ZM_{magPhase}LBP$	94.59	94.83	96.67	96.78	96.67
$ZM_{component}LBP$	90.15	92.67	93.91	95.22	96.67
$ZM_{mag}LTP$	92.96	94.75	96.38	97.44	98.13
$ZM_{magPhase}LTP$	93.11	95.5	96.38	97.0	97.6
$ZM_{component}LTP$	91.04	93.92	94.57	95.89	96.53

Table 6. Performance of the individual/combined approaches over illumination and expression variation on Yale database

Descriptor	YALE 3.1	YALE 3.2	YALE 4.1	YALE 4.2
ZM_{mag}	46.67	80.0	47.78	82.22
$ZM_{magPhase}$	61.67	84.0	54.44	81.11
$ZM_{component}$	53.33	81.33	50.0	81.11
LBP	83.33	85.33	66.67	96.67
LTP	76.67	76.0	72.22	96.67
$ZM_{mag}LBP$	91.67	97.33	84.44	97.78
$ZM_{magPhase}LBP$	88.33	96.0	85.56	98.89
$ZM_{component}LBP$	88.33	97.33	77.78	94.44
$ZM_{mag}LTP$	90.0	96.0	77.78	97.78
$ZM_{magPhase}LTP$	90.0	96.0	82.22	97.78
$ZM_{component}LTP$	88.33	94.67	72.22	96.67

For illumination variation, i.e., on YALE 3.1 category, the highest recognition rate of 91.67% is achieved by the proposed $ZM_{mag}LBP$ approach, whereas against expression variation, the $ZM_{magPhase}LBP$ descriptor gives the highest recognition rate at 85.56% on YALE 4.1 category. Experiments are also conducted on YALE 3.2 category wherein all of the face images consisting of expression variation are taken in the training set and the remaining ones, i.e., one neutral and four images with illumination changes, are placed in the test set. The results for this setup are also shown in Table 6, from which it is observed that the performance of the $ZM_{mag}LBP$ as well as that of $ZM_{component}LBP$ is better. Particularly, on YALE 3.2 category, a superior recognition rate of 97.33% is achieved by both approaches. Similarly, in case of YALE 4.2 category, four images of each person consisting of illumination variation are used to create the training set while all of the remaining ones (i.e., one neutral and six images in varying expressions) are placed in the test set. As shown in Table 6, $ZM_{magPhase}LBP$ achieves a high recognition rate of 98.89% for this category. Thus, from the results shown in Table 6, it can be concluded that $ZM_{mag}LBP$ is illumination invariant and $ZM_{magPhase}LBP$ is expression invariant. If we look at the overall performance of the proposed approaches on Yale database, $ZM_{mag}LBP$ and $ZM_{magPhase}LBP$ outperform the other combinations.

5.3 Performance on ORL database

The ORL face database consists of a total of 400 images of size 112 × 92 pixels of 40 persons with ten images per person in different states of variation. All the face images in this database are taken against a dark homogenous background. These images contain slight pose variation (tilt and yaw) up to ±20° with some basic facial expressions (smiling/not smiling, open/closed eyes). For performing experiments, the images of this database are cropped to 64 × 64 pixels. Sample cropped images for one person are shown in Figure 7. The face images of these databases are partitioned into 64 patches of 8 × 8 pixels to extract the local LBP/LTP features. Detailed experiments are performed on this database in order to analyze robustness of the proposed combined approaches against the pose variation. Various data partitions generated for this purpose are presented in Table 7.

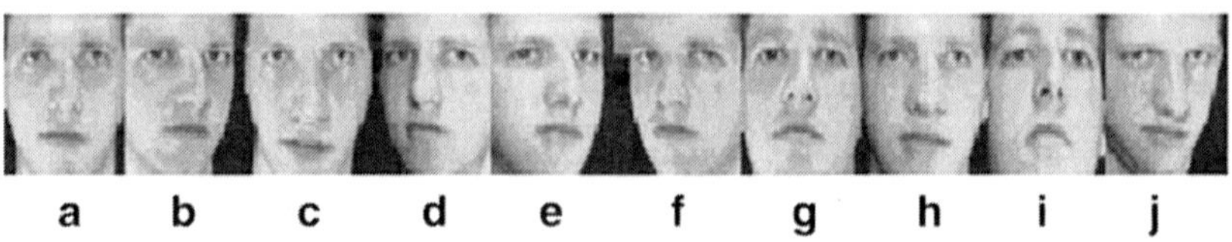

Figure 7. Sample cropped face images (a-j) for one person from ORL database.

Table 7. Data partition on ORL database for performing various experiments

Set of experiment	Category	Training	Testing	Remarks
1	ORL 1	Random one	All remaining (i.e., remaining ten images except the one selected for the training)	Ten different trials of this setup have been taken, and the recognition result (Figure 8) is the average of all these trials
2	ORL 2.1	Random two	Remaining eight	The recognition result (shown in Table 8) is the average of ten random trials on each setup, i.e., ORL 2.1, ORL 2.2, ORL 2.3, and ORL 2.4
	ORL 2.2	Random three	Remaining seven	
	ORL 2.3	Random four	Remaining six	
	ORL 2.4	Random five	Remaining five	
3	ORL 3	a, b	c, g, i, j	Testing consists of experiments against scale and up/down (tilt) pose variation
4	ORL 4	a, b	b, d, e, f	Testing consists of experiments against left/right (yaw) pose variation

Firstly, experiments are performed on ORL 1 category by taking one image of each person in the training set, and all of the remaining ones are used to formulate the test set. Different trials are framed in this case. As there are nine different images in the test set, ten combinations of different training and test images are possible here. The average recognition results over these ten different trials are shown in Figure 8a,b. The results on different values of n_{max} are depicted in order to analyze the effect of maximum order of moments n_{max} of ZMs on the performance of the proposed combined approaches. As the basic LBP and the LTP descriptors used in this work are not invariant to image rotation whereas the ZM descriptor is an established rotation invariant scheme, it is observed from the results that the performance of the individual ZM descriptors is better than that of the LBP/LTP descriptors for this database. Among the ZM-based descriptors, $ZM_{component}$ and $ZM_{magPhase}$ descriptors give the highest recognition rates because of the inclusion of phase coefficients. However, an improvement of more than 10% is achieved by fusion of the invariant feature sets of the ZM and LBP/LTP descriptors wherein the ZM descriptor plays a significant role in achieving rotation invariance. The highest recognition rate of 81.22% is achieved by the proposed ZM_{mag}LTP approach. From the results presented in Figure 8a,b, it is observed that in the proposed combined methods, fusion of the ZM features obtained at $n_{max} = 9$ provides better results on this database. Accordingly, further experiments have been conducted only on this order of moments.

The average recognition results over ten different trials of each group (i.e., ORL 2.1, ORL 2.2, ORL 2.3, and ORL 2.4) of the training and the test sets are presented in Table 8. Excellent results are obtained by the proposed combined approaches, while $ZM_{component}$LTP provides the best results. On taking five images in training and the remaining five in the test set (i.e., for ORL 2.4 over ten runs), the average recognition rate of 99.2% is achieved by both the ZM_{mag}LBP and $ZM_{component}$LTP approaches. Further, $ZM_{component}$ features have proven to be invariant to image rotation and tolerant to pose variations to some extent [29]. From this analysis, we can state that the $ZM_{component}$ combined with LTP as well as the ZM_{mag} coupled with LBP provides superior results against pose variations.

Next, experiments are performed on ORL 3 category by taking two neutral face images in the training set, while four images of each person consisting of scale and up/down head movement are taken in the test set. Similarly, in order to examine the performance of the proposed approaches over yaw pose variation, two neutral images of each person are placed in the training set and four images with slight left/right head movement are placed in the test set, i.e., ORL 4 category. The results obtained from this experimental analysis are also presented in Table 8. On ORL 3 category, the performance of the $ZM_{component}$ coupled with that of the LBP/LTP descriptors is better, achieving a recognition rate of 90.0%. Similarly, on ORL 4 category, the highest recognition rate of 91.25% is achieved by both the $ZM_{mag}LBP$ and the $ZM_{magPhase}LBP$ approaches. Thus, in most of the cases of ORL database, the $ZM_{component}$combined with LBP/LTP outperforms the other proposed combinations.

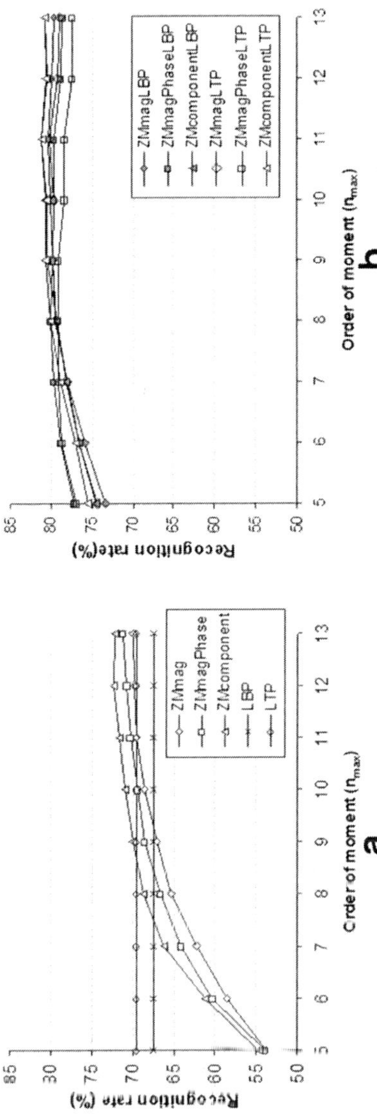

Figure 8. Performance (average) on ORL 1 category. Recognition results of **(a)** individual approaches and **(b)**combined approaches.

Table 8. Performance of the considered approaches against different pose variations on ORL database

Descriptor	ORL 2.1	ORL 2.2	ORL 2.3	ORL 2.4	ORL 3	ORL 4
ZM_{mag}	81.87	88.71	93.08	94.55	81.88	80.63
$ZM_{magPhase}$	83.5	89.18	94.21	95.3	85.63	81.88
$ZM_{component}$	84.81	90.82	94.96	95.95	84.38	83.75
LBP	82.23	88.75	92.92	95.7	80.0	84.38
LTP	83.28	89.14	92.21	95.4	82.5	86.25
ZM_{mag}LBP	91.24	94.96	97.5	*99.2*	88.75	*91.25*
$ZM_{magPhase}$LBP	91.34	94.57	96.71	98.25	88.75	*91.25*
$ZM_{component}$LBP	*92.47*	95.07	97.5	98.9	*90.0*	88.75
ZM_{mag}LTP	90.69	94.39	97.08	98.85	88.13	88.13
$ZM_{magPhase}$LTP	90.0	93.64	96.5	98.35	86.88	90.0
$ZM_{component}$LTP	92.17	*95.21*	*97.79*	*99.2*	*90.0*	88.13

5.4 Performance analysis against noise variation

To examine the effect of noise on the recognition accuracy, we add impulsive noise, commonly named salt-and-pepper or spike noise, to the face images of the three databases. In the presence of impulsive noise, an image has dark pixels in bright regions and white pixels in dark regions [30]. In this analysis, a noise of 0.05 is added to the images of the test set whereas the training is done on original face images, i.e., on images with no noise. The procedure of experimental setup to examine the performance of these approaches against additive noise is the same as before. That is, in order to analyze the performance on FERET database, experiments are performed on FERET_A3 data partition in which one frontal image (0° pose) is selected in the training set and the four images in different poses (±22.5°, ±67.5° and with additive noise) for each person are used in the test set. On YALE 2.4 data partition, robustness of the proposed approaches is analyzed against noise variation by selecting five images of each person in the training set and the remaining six images (with additive noise) in the test set. The results presented are the average recognition rates over ten different runs of training and test sets. In a similar manner, the experiments on the images of ORL 2.4 data partition are performed by taking random five images of each person in the training set and the remaining five images (with additive noise) in the test set, and the recognition results for the same are also the averaged recognition rates over ten different runs of training and test sets. The experimental results on the said databases are shown in Table 9.

From the results presented, it is observed that among the individual approaches, the LTP descriptor is more robust to noise variation than the LBP. On Yale and ORL databases (with noise variation), the performance of the proposed $ZM_{mag}LTP$ and $ZM_{component}LTP$ descriptors, respectively, is better as compared to all other combined approaches. On the other hand, if FERET images with noise variations are assessed, then the recognition rate of $ZM_{magPhase}LBP$ is 81.5%, whereas the recognition rate of $ZM_{mag}LTP$ is 81.0%. The percentage difference between the actual results obtained (without adding noise and with noise) for both approaches is 5 and 3.5%, respectively. Hence, from this observation, we can say that on FERET database, the performance of the proposed $ZM_{mag}LTP$ descriptor is better against noise variation. For the case of Yale and ORL databases, the degradation due to noise in recognition rates is very less.

Table 9. Performance of the considered approaches against noise variation

Descriptor	Face datasets comprising additive noise in testing		
	FERET_A3	YALE 2.4	ORL 2.4
ZM_{mag}	62.75	80.22	94.65
$ZM_{magPhase}$	65.75	75.33	94.45
$ZM_{component}$	64.75	80.11	95.55
LBP	64.75	91.0	95.45
LTP	67.75	94.0	95.85
ZM_{mag}LBP	80.0	96.22	98.55
$ZM_{magPhase}$LBP	*81.5*	94.56	98.0
$ZM_{component}$LBP	77.0	93.33	98.85
ZM_{mag}LTP	*81.0*	*96.44*	98.85
$ZM_{magPhase}$LTP	79.25	96.11	98.45
$ZM_{component}$LTP	79.0	94.33	*99.0*

5.5 Time complexity

One of the important issues involved in using combined approaches similar to the ones proposed here is the time complexity of these approaches. It is a common perception that the moment-based descriptors are computation intensive which is true to some extent especially in case of the ZM calculation. The time complexity of the ZMs is of order $O(N^2 n_{max}^3)$ if all moments up to a maximum order n_{max} are computed for an image of $N \times N$ pixels. However, with the use of fast algorithms [10, 11], the time complexity is reduced to $O(N^2 n_{max}^2)$. Further significant reduction in computation time is achieved by using symmetry/antisymmetry properties of kernel function of ZMs. The ZMs of the database images are computed offline and indexed with the images themselves. The ZMs of the test image are computed online. Although the time complexity of ZM calculation is still high, in this work, better recognition results have been obtained with $n_{max} = 9$ for FERET and ORL databases whereas Yale database exhibits good results by taking $n_{max} = 11$; therefore, we consider moments only up to these orders. As $Z_{0,0}$ and $Z_{1,1}$have no discriminative capabilities, they do not affect the recognition rate. Hence, with $n_{max} = 11$, we have 40 features after discarding the coefficients $Z_{0,0}$ and $Z_{1,1}$. In contrast, although the number of features in the feature vector containing local histogram features of the LBP/LTP descriptors is high, the computation time of these descriptors is very low. Thus, the proposed fusion of the ZM and LBP/LTP descriptors maintains a good balance between speed and dimensionality. The size of feature vectors of the ZM and LBP/LTP descriptors is shown in Table 10.

Table 10. Dimensionality of the feature vectors of the ZM and LBP/LTP descriptors

Approach	Size of the feature vector	Remarks
ZM_{mag}	40	For moments up to order $n_{max} = 11$
$ZM_{magPhase}$	2×40	Size of the feature vector is double the size of magnitude features
$ZM_{component}$	2×40	Size of the feature vector is double the size of magnitude features
LBP	3,776	Taking image patch size of 8×8 pixels for uniform local binary patterns
LTP	$2 \times 3{,}776$	Number of features is double the number of LBP features because two feature vectors consisting of the positive and the negative uniform binary patterns are taken

We observe that for an image of 256×256 pixels, the CPU elapse time for calculating ZMs is only 0.032 s for $n_{max} = 12$ on a PC with a 3.0-GHz CPU and a 1-GB RAM under Microsoft Windows environment. The time taken for computing LBP and LTP features is 0.015 and 0.016 s, respectively. Thus, the total time elapsed for the extraction of the local and global features of a test image does not exceed 0.048 s. The time taken for classification is much less than the feature extraction time. Thus, in comparison to the gain in the recognition performance, the time taken by the combined features is much less and can be afforded by the low computation power devices in online mode. Since the time complexity does not depend on the contents of the image, these experiments are carried out for one image only.

5.6 Performance comparison

We have compared the performance of the proposed combined descriptors with other popular methods such as PCA, 2DPCA, $(PC)^2A$, $E(PC)^2A$, $2D(PC)^2A$, SVD perturbation [38], and hybrid Fourier-AFMT transform [39] for face recognition with single (first) example image per person. As shown in Table 11, the proposed combined descriptors give the best recognition rate when compared with other well-established methods. On the other hand, the time complexity of PCA-based methods is very high as compared to the proposed approaches.

Table 11. Performance comparison (%) of some recent approaches with proposed methods on Yale and ORL databases

Method	ORL	Yale
PCA	55.0	55.0
2DPCA	54.0	56.0
$(PC)^2A$	57.0	55.0
$E(PC)^2A$	58.0	57.0
$2D(PC)^2A$	56.0	55.0
SVD perturbation	61.0	62.0
Hybrid Fourier-AFMT transform	76.0	52.0
RES [40]	0.77	0.78
$ZM_{mag}LBP$	*84.44*	*86.67*
$ZM_{magPhase}LBP$	84.17	*86.67*
$ZM_{component}LBP$	84.17	80.67
$ZM_{mag}LTP$	80.0	82.67
$ZM_{magPhase}LTP$	82.78	85.33
$ZM_{component}LTP$	81.22	81.52

Comparison of performance of the proposed combined descriptors with other popular methods for face recognition with single (first) example image per person.

Dual optimal multiband features (DOMF) [24] give a recognition rate of 92.6 and 88.4% on Yale and ORL databases, respectively, when two images of each person are taken in the training set and all the remaining are kept in the test set. On this similar setup for training and test images, the highest recognition rate achieved by the proposed $ZM_{magPhase}$LBP descriptor for YALE 2.1 is 94.59% while the $ZM_{component}$LBP descriptor achieves a recognition rate of 92.47% for ORL 2.1 category.

The performance of the proposed combined approaches is also compared with that of some recent face recognition methods when five images of each person are used for training. The recognition results of the proposed combined approaches and those of these recent methods on Yale and ORL databases for this case are shown in Table 12. The best results are highlighted in italics. All these methods use multidimensional features or combined approaches to represent the face images. As can be seen from the results presented, the recognition rate of the proposed approaches is higher as compared to that of the recent methods. In case of block-based S-P approach [25], one random set of five images per person is taken in the training set while all the remaining are kept in the test set for both the Yale and ORL databases, whereas the results presented for our proposed approaches are the average of ten random trials of training and test sets. It is worth mentioning here that on some of the random trials, our proposed descriptors also provide 100% recognition rate. Recently introduced wavelet moment (WM) and complex WM (CWM) approaches [41] have achieved a recognition rate of 51.5 and 54.3%, respectively, on FERET_A2 subset, while the proposed $ZM_{magPhase}$LBP descriptor has attained a recognition rate of 77.33%. On the *fafb* subset of FERET database, the recognition rate obtained by the RES [40], WM, and CWM [41] approaches is 95.0, 88.0, and 91.0% whereas the highest recognition rate achieved by the proposed $ZM_{magPhase}$LBP approach is 98.04%. Thus, on the basis of superior results obtained by the proposed fusion technique, it can be concluded that combining the feature sets of the ZM and LBP/LTP descriptors is an efficient and practical approach for robust face recognition.

Table 12. Performance comparison (%) of the proposed approaches with recent methods on Yale and ORL databases

Method	Yale database	ORL database
Two-dimensional LDA (2D-LDA) [42]	86.57	92.50
2D-WLDA [42]	88.00	93.50
2D-DWLDA [42]	89.33	94.00
Direct LDA (DLDA) [21]	93.20	92.50
Enhanced Fisher linear discriminant model (EFM) [21]	93.90	92.50
Intrinsicfaces [43]	74.00	97.00
Combined feature Fisher classifier (CF^2C) [21]	96.90	96.80
Feature Fisher classifier (F^3C) [20]	96.4	94.9
Block based S-P[a] [25]	100.0[a]	99.0
Algorithm A (WMs) [41]	–	93.5
Algorithm B (CWMs) [41]	–	96.0
$ZM_{mag}LBP$	*97.56*	*99.20*
$ZM_{magPhase}LBP$	96.78	98.25
$ZM_{component}LBP$	95.67	98.95
$ZM_{mag}LTP$	*97.56*	98.85
$ZM_{magPhase}LTP$	97.00	98.35
$ZM_{component}LTP$	97.00	*99.20*

Comparison of recognition performance for five random training images. [a]The results are presented on only one random set of five images in the training set and all the remaining in the test set.

6 CONCLUSIONS

This paper proposes the fusion of two useful feature sets, i.e., the global ZMs and the local LBP/LTP descriptor. Face images capture extensive variation under varying pose and lighting conditions accompanied by the presence of expression and noise. Individually, the ZM and LBP/LTP descriptors are observed to be very effective in providing good recognition performance on the face images containing certain variations. In particular, the ZM descriptor extracts rotationally invariant shape features from the whole face images, whereas the LBP/LTP descriptors are able to capture the fine details and illumination-invariant characteristics within some local regions of the face images. However, the fusion of these two complementary approaches incorporates the benefits of both of these descriptors and as such proves to be invariant against various distortions present in the face images. Herein this work, diverse feature sets of ZMs are combined with LBP/LTP descriptors to generate various combined approaches, namely, $ZM_{mag}LBP$, $ZM_{magPhase}LBP$, $ZM_{component}LBP$, $ZM_{mag}LTP$, $ZM_{magPhase}LTP$, and $ZM_{component}LTP$. From the detailed experiments performed on FERET, Yale, and ORL face databases, it has been observed that the proposed combined approaches are highly robust against pose, expression, illumination, and noise variations, as the recognition rate achieved by the proposed approaches is approximately 10 to 30% higher than that obtained by applying these approaches individually. Fusion of ZM and LBP descriptor performs better over the pose, expression, and illumination variations, while in the presence of noise, ZMs combined with LTP descriptor generate superior results. Experimental results also prove the efficacy of the proposed methods over other existing techniques. Also, significant improvement in the recognition rate is achieved by the proposed scheme when only single training image per person is available.

Future work is suggested towards discovering the optimal ways to utilize the information acquired by the phase coefficients of ZM descriptor in addition to using different methods of classification to further improve the performance of the proposed fusion approach.

ACKNOWLEDGEMENTS

The authors are thankful to the useful comments and suggestions of the anonymous reviewers for raising the standard of the paper. The authors are grateful to the All India Council for Technical Education (AICTE), Govt. of India, New Delhi, India, for supporting the research work vide their file number 8013/RID/BOR/RPS-77/2005-06. We are also grateful to the National Institute of Standards and Technology (colorferet@nist.gov) for providing FERET face database.

COMPETING INTERESTS

The authors declare that they have no competing interests.

REFERENCES

1. Zhao W, Chellappa R, Phillips P, Rosenfeld A: Face recognition: a literature survey. *ACM Comput. Surv.* 2003, 35(4):399-458.

2. Hjelmas E, Low BK: Face detection: a survey. *Comput. Vision. Image. Underst.* 2001, 83: 236-274.

3. Turk M: A random walk through Eigenspace. *IEICE Trans. Inf. Syst.* 2001, E84-D(12):1586-1595.

4. Mittal N, Walia E: Face recognition using improved fast PCA algorithm. In *Proceedings of the IEEE International Congress on Image and Signal Processing (CISp '08)*. Sanya, Hainan; 2008:554-558.

5. Belhumeur PN, Hespanha JP, Kriegman DJ: Eigenfaces vs. Fisherfaces: recognition using class specific linear projection. *IEEE. IEEE Trans. Pattern Anal. Mach. Intell.* 1997, 19: 711-720.

6. Xu Y, Zhang D, Yang J, J–Y Y: An approach for directly extracting features from matrix data and its application in face recognition. *Neurocomputing* 2008, 71: 1857-1865.

7. Daoqiang Z, Zhi-Hua Z: $(2D)^2$PCA: Two-directional two-dimensional PCA for efficient face representation and recognition. *Neurocomputing* 2005, 69: 224-231.

8. Zhang D, Lu G: Review of shape representation and description techniques. *Pattern Recognit.* 2004, 37(1):1-19.

9. Zhang D, Lu G: Evaluation of MPEG-7 shape descriptors against other shape descriptors. *Multimed. Syst.* 2003, 9: 15-30.

10. Singh C, Walia E: Fast and numerically stable methods for the computation of Zernike moments. *Pattern Recognit.* 2010, 43(7):2497-2506.

11. C–Y W, Paramesran R: On the computational aspects of Zernike moments. *Image Vis. Comput.* 2007, 25: 967-980.

12. Lowe DG: Distinctive image features from scale-invariant keypoints. *Int. J. Comput. Vis.* 2004, 60(2):91-110

13. Soyel H, Demirel H: Facial expression recognition based on discriminative scale invariant feature transform. *IET Electron. Lett.* 2010, 46(5):343-345.

14. Huang L, Shimizu A, Kobatake H: Robust face detection using Gabor filter features. *Pattern Recognit. Lett.*2005, 26(11):1641-1649.

15. Ojala T, Pietikäinen M: Multiresolution gray-scale and rotation invariant texture classification with local binary patterns. *IEEE Trans. Pattern Anal. Mach. Intell.* 2002, 24(7):971-987.

16. Ahonen T, Hadid A, Pietikäinen M: Face description with local binary patterns: application to face recognition.*IEEE Trans. Pattern Anal. Mach. Intell.* 2006, 28(12):2037-2041.

17. Jun B, Kim T, Kim D: A compact local binary pattern using maximization of mutual information for face analysis.*Pattern Recognit.* 2011, 44: 532-543.

18. Tan X, Triggs B: Enhanced local texture feature sets for face recognition under difficult lighting conditions. *IEEE Trans. Image Process.* 2010, 19(6):1635-1648

19. Kim C, Oh J, Choi C: *Combined Subspace Method Using Global and Local features For Face Recognition*. 4th edition. Proceedings of the IEEE International Joint Conference on Neural Networks (IJCNN '05), Montreal Canada; 2005:2030-2035.

20. Zhou D, Yang X: Feature fusion based face recognition using EFM. Proceedings International Conference on Image Analysis and Recognition (ICIAR '04). *Lect. Notes Comput. Sci.* 2004, 3212: 643-650.

21. Zhou D, Yang X, Peng N, Wang Y: Improved LDA based face recognition using both facial global and local information. *Pattern Recognit. Lett.* 2006, 27: 536-543.

22. Singh C, Walia E, Mittal N: Robust two-stage face recognition approach using global and local features. *Vis. Comput.* 2012, 28(11):1085-1098.

23. Su Y, Shan S, Chen X, Gao W: Hierarchical ensemble of global and local classifiers for face recognition. *IEEE Trans. Image Process.* 2009, 18(8):1885-1895.

24. Wong Y-W, Seng KP, Li-M A: Dual optimal multiband features for face recognition. *Expert Syst, Appl* 2010, 37(4):2957-2962.

25. Aroussi ME, Hassouni ME, Ghouzali S, Rziza M, Aboutajdine D: Local appearance based face recognition method using block based steerable pyramid transform. *Signal Process* 2011, 91: 38-50.

26. Liu Z, Liu C: Fusion of color, local spatial and global frequency information for face recognition. *Pattern Recognit.* 2010, 43: 2882-2890.

27. Jun B, Lee J, Kim D: A novel illumination-robust face recognition using statistical and non-statistical method.*Pattern Recognit. Lett.* 2011, 32: 329-336.

28. Moore S, Bowden R: Local binary patterns for multi-view facial expression recognition. *Comput. Vision Image Underst.* 2011, 115: 541-558.

29. Singh C, Walia E, Mittal N: Rotation invariant complex Zernike moments features and their application to human face and character recognition. *IET Comput. Vision* 2011, 5(5):255-265.

30. Lajevardi SM, Hussain ZM: Higher order orthogonal moments for invariant facial expression recognition. *Digit Signal Process* 2010, 20: 1771-1779.

31. Li S, M–C L, Chi-Man P: Complex Zernike moments features for shape based image retrieval. *IEEE Trans. Syst. Man. Cybern. C Appl. Rev.* 2009, 39: 227-237.

32. Singh C, Mittal N, Walia E: Face recognition using Zernike and complex Zernike moment features. *Pattern Recognit. Image Anal.* 2011, 21(1):71-81.

33. Revaud J, Lavoue G, Baskurt A: Improving Zernike moments comparison for optimal similarity and rotation angle retrieval. *IEEE Trans. Pattern Anal. Mach. Intell.* 2009, 31(4):627-636.

34. Jain A, Nandakumar K, Ross A: Score normalization in multimodal biometric systems. *Pattern Recognit.* 2005, 38: 2270-2285.

35. The Facial Recognition Technology (FERET) face database http://www.nist.gov/itl/iad/ig/colorferet.cfm

36. Yale face database http://www.cl.cam.ac.uk/research/dtg/attarchive/facedatabase.html

37. Olivetti Research Laboratory (ORL) face database http://www.cl.cam.ac.uk/research/dtg/attarchive/facedatabase.html

38. Li J, Pan J-S: A novel pose and illumination robust face recognition with a single training image per person algorithm. *Chin. Optic. Lett.* 2008, 6(4):255-257.

39. Chen YM, Chiang J-H: Fusing multiple features for Fourier Mellin-based face recognition with single example image per person. *Neurocomputing* 2010, 73(16–18):3089-3096.

40. Kuo C-H, Lee JD: Face recognition based on a two-view projective transformation using one sample per subject. *IET Comput. Vision* 2012, 6(5):489-498.

41. Singh C, Sahan AM: Face recognition using complex wavelet moments. *Opt. Laser Technol.* 2013, 47: 256-267.

42. Zhi R, Ruan Q: Two-dimensional direct and weighted linear discriminant analysis for face recognition.*Neurocomputing* 2008, 71: 3607-3611.

43. Wang Y, Wu Y: Face recognition using Intrinsicfaces. *Pattern Recognit.* 2010, 43: 3580-3590.

Chapter 6

3D Face Recognition

Zaeri Naser

[1] Arab Open University, Kuwait

1. INTRODUCTION

Biometric systems for human recognition are an ongoing demand. Among all biometric technologies which are employed so far, face recognition is one of the most widely outspread biometrics. Its daily use by nearly everyone as the primary mean for recognizing other humans and its naturalness have turned face recognition into a well-accepted method. Furthermore, this image procurement is not considered as intrusive as the other mentioned alternatives.

Nonetheless, in spite of the various facial recognition systems which already exist, many of them have been unsuccessful in matching up to expectations. 2D facial recognition systems are constrained by limitations such as physical appearance changes, aging factor, pose and changes in lighting intensity. Recently, to overcome these challenges 3D facial recognition systems have been issued as the newly emerged biometric technique, showing a high level of accuracy and reliability, being more robust to face variation due to the different factors.

A face-based biometric system consists of acquisition devices, preprocessing, feature extraction, data storage and a comparator. An acquisition device maybe a 2D-, 3D- or an infra-red- camera that can record the facial information. The preprocessing can detect facial landmarks, align facial data and crop facial area. It can filter irrelevant information such as hair, background and reduce facial variation due to pose change. In 2D images, landmarks such as eye, eyebrow, mouths etc, can be reliably detected, in contrast, nose is the most important landmark in 3D face recognition.

The 3D information (depth and texture maps) corresponding to the surface of the face may be acquired using different alternatives: A multi camera system (stereoscopy), range cameras or 3D laser and scanner devices. Different approaches have been presented from the 3D perspective. The first approach would correspond to all 3D approaches that require the same data format in the training and in the testing stage. The second philosophy would enclose all approaches that take advantage of the 3D data during the training stage but then use 2D data in the recognition stage. Approaches of the first category report better results than of the second group; however, the main drawback of this category is that the acquisition conditions and elements of the test scenario should be well synchronized and controlled in order to acquire accurate 3D data. Thus, they are not suitable for surveillance applications or control access points where only one “normal” 2D texture image (from any view) acquired from a single camera is available. The second category encloses model-based approaches. Nevertheless, model-based face recognition approaches present the main drawback of a high computational burden required to fit the images to the 3D models.

In this chapter, we study 3D face recognition where we provide a description of the most recent 3D based face recognition techniques and try to coarsely classify them into categories, as explained in the following subsequent sections.

2. ITERATIVE CLOSEST POINT

(Maurer et al., 2005) presented a multimodal algorithm that uses Iterative Closest Point (ICP) to extract distance map, which is the distance between mesh of reference and probe. This method includes, face finding, landmark finding, and template computation. They used weighted sum rule to fuse shape and texture scores. If 3D score is high, algorithm uses only shape for evaluation. In experimental tests by using 4007 faces in the FRGC v2 database, a verification rate of 87.0% was achieved at %0.1 false accept rate (FAR). (Kakadiaris et al., 2007)performed face recognition with an annotated model that is non-rigidly registered to face meshes through a combination of ICP, simulated annealing and elastically adapted deformable model fitting. A limitation of this approach is the imposed constraints on the initial orientation of the face.

Performance of 3D methods highly depends on registration performance, where ICP is commonly used. ICP registration performance is highly dependent on initial alignment and it performs solid registration. However, expression variations degrade registration success. To overcome this problem, (Faltemier et al., 2008) divided the face into different overlapping regions where each face region was registered independently. Distance between regions was used as a similarity measure and results were fused using modified Borda count. They achieved 97.2% rate on FRGC v2 database. Other approaches to discard the effect of expressions were also studied by dividing the face into separate parts and extracting features from each part in 2D and range images (Cook et al., 2006; McCool et al., 2008).

3. GEOMETRIC APPROACH

The early work of applying invariant functions on 3-D face recognition was done over a decade ago. At that time, people began with the geometrical properties introduced in differential geometry, such as principal curvatures, Gaussian curvature, etc. Basically, these approaches use the invariant functions, e.g., Gaussian curvature which is invariant under Euclidean transformations, to extract information from the face surface and, then, perform a classification that is based on the extracted information. (Riccio & Dugelay, 2007) proposed a particular 2D-3D face recognition method based on 16 geometric invariants, which were calculated from a number of "control points". The 2D face images and 3D face data are related through those geometric invariants. The method is invariant to pose and illumination, but the performance of the method closely depend on the accuracy of "control points" localization.

In the approach proposed by (Elyan & Ugail, 2009), the first goal was to automatically determine the symmetry profile along the face. This was undertaken by means of computing the intersection between the symmetry plane and the facial mesh, resulting in a planner curve that accurately represents the symmetry profile. Once the symmetry profile is successfully determined, a few feature points along the symmetry profile are computed. These feature points are essential to compute other facial features, which can then be utilized to allocate the central region of the face and extract a set of profiles from that region (Fig. 1). In order to allocate the symmetry profile, it was assumed that it passes through the tip of the nose. This was considered as the easiest feature point to recover and to allocate using a bilinear blended Coon's surface patch. Coon's patch is a parametric surface defined by a given four-boundary curves. In (Elyan & Ugail, 2009), the four boundaries of the coon's patch were determined based on a boundary curve that encloses an approximated central region of interest, which is simply the region of the face that contains or likely to contain the nose area. This region was approximated based on the centre of the mass that represents the 3D facial image. They have computed the Fourier coefficients of the designated profiles and stored it in a database, other than storing the actual points of the profile. Thus, having a database of images representing different individuals where each person was represented by two profiles stored by means of their Fourier coefficients.

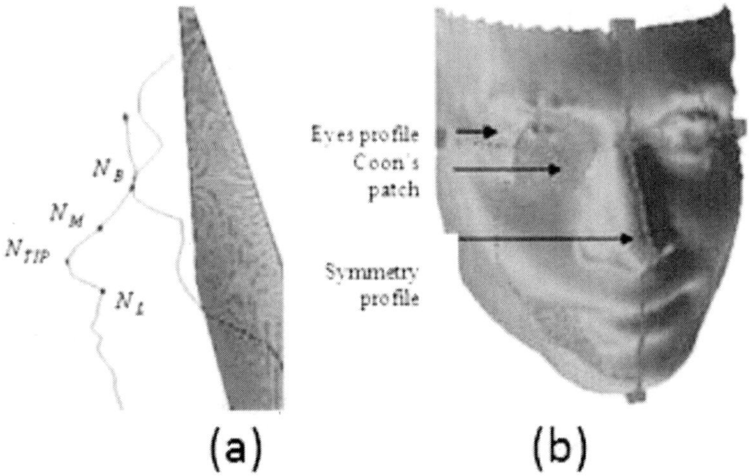

Figure 1. Facial features identification (a) Symmetry profile identification and analysis based on depth value to the reference depth plane (b) Eyes profile shown as the profile that passes through the nose bridge (Elyan & Ugail, 2009)

Moreover, several works in the literature propose to map 3D face models into some low-dimensional space, including the local isometric facial representation (Bronstein et al., 2007), or conformal mapping (Wang et al., 2006). Some works, for simplification, try also to investigate partial human biometry, meaning recognition based only on part of a face, as for example in (Drira et al., 2009), where authors used the nose region for identification purposes. (Szeptycki et al., 2010) explored how conformal mapping to 2D space (Wang et al., 2006) can be applied to partial face recognition. To deal with the computational cost of 3D face recognition they have utilized conformal maps of 3D surface to a 2D domain, thus simplifying the 3D mapping to a 2D one. The principal issue addressed in (Szeptycki et al., 2010) was to create facial feature maps which can be used for recognition by applying previously developed 2D recognition techniques. Creation of 2D maps from 3D face surfaces can handle model rotation and translation. However, their technique can be applied only to images with variation in pose and lighting. The expression changes were avoided. To create face maps which are later used for recognition, they started with models preprocessing (hole, spike removal). Next step was to segment the rigid part of a face that has less potential to change during expression. Finally, they performed UV conformal parameterization as well as shape index calculation for every vertex; the process is shown in Fig. 2.

(Song et al., 2009) detected the characteristics of the three regions eyes, nose and mouth in the human face, and then calculated the geometric characteristics of these regions by finding the straight-line Euclidean distance, curvaturedistance, area, angle and volume.

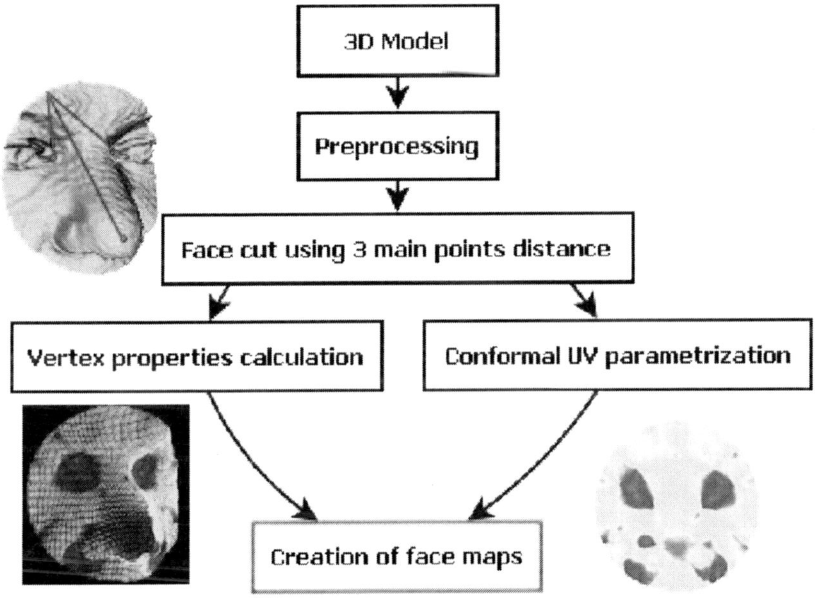

Figure 2. Face maps creation flow chart (Szeptycki et al., 2010)

Another face recognition system that is based on 3D geometric features was developed by (Tin & Sein, 2009). It is based on the perspective projection of a triangle constructed from three nodal points extracted from the two eyes and lips corners (Fig. 3). The set of non-linear equations was established using the nodal points of a triangle built by any three points in a 2D scene.

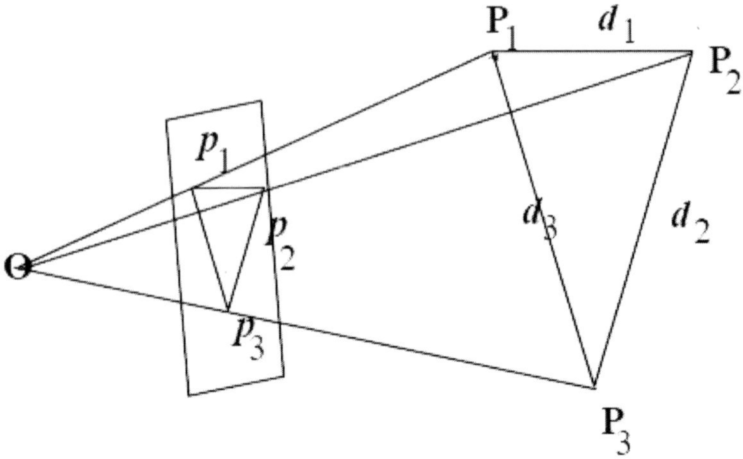

Figure 3. Illustration of Perspective Projection of a 3D Triangle (Tin & Sein, 2009)

An automatic 3D face recognition system using geometric invariant feature was proposed by (Guo et al., 2009). They utilized two kinds of features, one is the angle between neighboured facets, they made it as the spatial geometric feature; the other is the local shape representation vector, and they made it as the local variation feature. They combined these two kinds of features together, and obtained the geometric invariant feature. Before feature extraction, they have presented a regularization method to build the regular mesh models. The angle between neighboured facets is invariant to scale and pose; meanwhile, local shape feature represents the exclusive individual shape.

(Passalis et al., 2007) focused on intra-class object retrieval problems, specifically, on face recognition. By considering the human face as a class of objects, the task of verifying a person's identity can be expressed as an intra-class retrieval operation. The fundamental idea behind their method is to convert raw polygons in R^3 space into a compact 2D description that retains the geometry information, and then perform the retrieval operation in R^2 space. This offers two advantages: 1)working in R^2 space is easier, and 2)the system can apply the existing 2D techniques. A 3D model is first created to describe the selected class. Apart from the geometry, the model also includes any additional features that characterise the class (e.g., area annotation, landmarks). Additionally, the model has a regularly sampled mapping from R^3 to R^2 (UV parameterization) that can be used to construct the equivalent 2D description, the geometry image. Subsequently, a subdivision-based model is fitted onto all the objects of the class using a deformable model framework. The result is converted to a geometry image and wavelet decomposition is applied. The wavelet coefficients are stored for matching and retrieval purposes (Fig. 4).

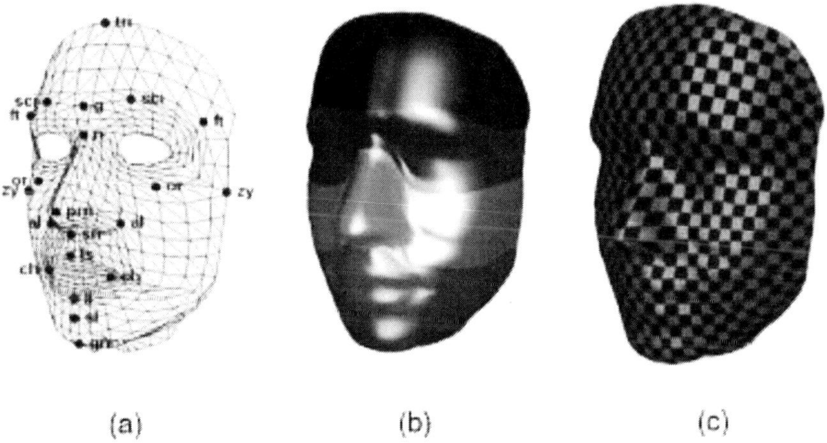

Figure 4. a) Anthropometric landmarks used, (b) segmentation into annotated areas, and (c) checkerboard texture to demonstrate parameterization (Passalis et al., 2007)

(Zaeri, 2011) investigated a new 3D face image acquisition and capturing system, where a test-bed for 3D face image feature characteristic and extraction was demonstrated.(Wong et al., 2007) proposed a multi-region face recognition algorithm for 3D face recognition. They identified the multiple sub-regions over a given range facial image and extracted summation invariant features from each sub-region. For each sub-region and the corresponding summation invariant feature, a matching score was calculated. Then, a linear fusion method was developed to combine the matching scores of individual regions to arrive at a final matching score. (Samir et al., 2006) described the face surface using contour lines or iso-contours of the depth function while using the nose tip as a reference point for alignment. The face surface is represented as a 2D image (e.g., depth-map), and then a 2D image classification techniques are applied. This approach requires that the surfaces are aligned by the iterative closest point algorithm or by feature-based techniques. Then, the deformable parts of the face are detected and excluded from the matching stage or downgrade their contribution during matching.

This, however, may lead to loss of information (e.g., excluding the lower part of the face) which is important for classification. A different approach is to use an active appearance model or in the general case, a 3D deformable model which may be fitted to the face surface. The difficulty in this case is in building a (usually linear) model that can capture all possible degrees of freedom hidden in facial expressions and fitting the model to the surface in hand.

The approach of (Mpiperis et al., 2007) relies on the assumption that the face is approximately isometric, which means that geodesic distances among points on the surface are preserved, and tries to establish an expression-invariant representation of the face. This technique does not have the disadvantages outlined in some other methods (loss of information and dealing with face variability). (Mpiperis et al., 2007) have considered the face surface as a 2D manifold embedded in the 3D Euclidean space, characterized by a Riemannian metric and described by intrinsic properties, namely geodesics (Figures 5 and 6).

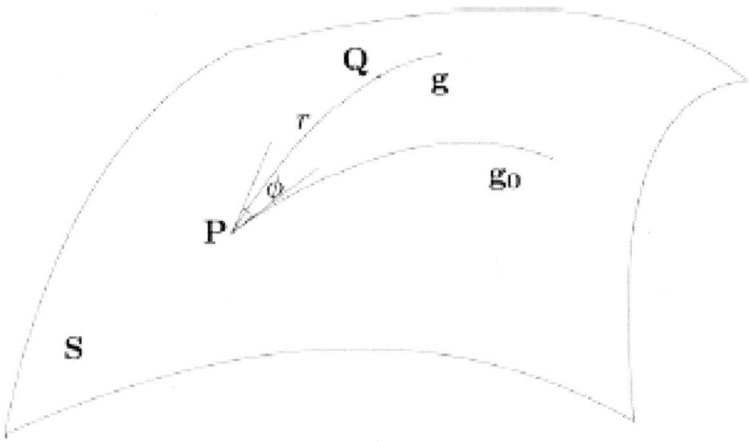

Figure 5. Definition of geodesic distance r and polar angle ϕ of an arbitrary point Q. Geodesic path g is the minimum length curve connecting point Q and geodesic pole P. r is the length of g, while ϕ is the angle between g and a reference geodesic path g0 (Mpiperis et al., 2007)

(Li et al., 2009) proposed a 3D face recognition approach using Harmonic Mapping and ABF++ as the mesh parameterization techniques. This approach represents the face surface in both local and global manners, which encodes the intrinsic attributes of surface in planar regions. Therefore, surface coarse registration and matching can be dealt with in a low dimensional space. The basic idea is to map 3D surface patches to a 2D parameterization domain and encode the shape and texture information of a 3D surface into a 2D image. Therefore, complex geometric processing can be analyzed and calculated in a low-dimensional space. The mean curvature to characterize the points of surface is employed. Then, both local shape description and global shape description with curvature texture are constructed to represent the surface. With the selected surface patches in local regions, Harmonic Mapping is used to construct the local shape description. Harmonic Mappings are the solutions to partial differential equations from the Dirichlet energy defined in Riemann manifolds. An example of the constructed local shape description at a feature point on a facial surface is shown in Fig. 7, while the global shape description is shown in Fig. 8. For the overall meshes of probe or gallery images, nonlinear parameterization ABF++ with free boundary, proposed by (Sheffer et al., 2005), is used to create global shape description.

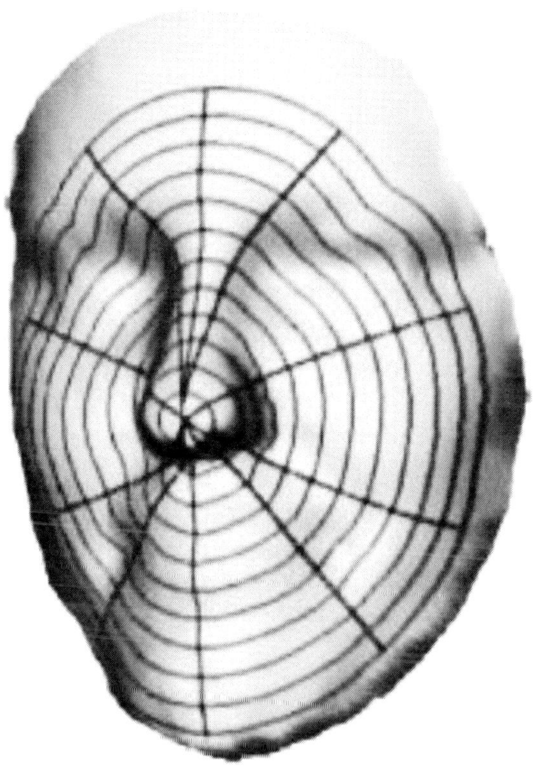

Figure 6. Geodesic paths and circles defined over a face surface. The tip of the nose was selected as the geodesic pole (Mpiperis et al., 2007)

The method presented by (Guo et al., 2010) is based on conformal geometric maps which does not need 3D models registration, and also maps 3D facial shape to a 2D domain which is a diffeomorphism through a global optimization. The 2D maps integrate geometric and appearance information and have the ability to describe the intrinsic shape of the 3D facial model, called Intrinsic Shape Description Maps (Fig. 9).

(Harguess & Aggarwal, 2009) presented a comparison of the use of the average-half-face to the use of the original full face with 6 different algorithms applied to two- and three- dimensional (2D and 3D) databases. The average-half-face is constructed from the full frontal face image in two steps; first the face image is centred and divided in half and then the two halves are averaged together (reversing the columns of one of the halves). The resulting average-half-face is then used as the input for face recognition algorithms. (Harguess & Aggarwal, 2009) compared the results using the following algorithms: eigenfaces, multi-linear principal components analysis (MPCA), MPCA with linear discriminant analysis (MPCALDA), Fisherfaces (LDA), independent component analysis (ICA), and support vector machines (SVM).

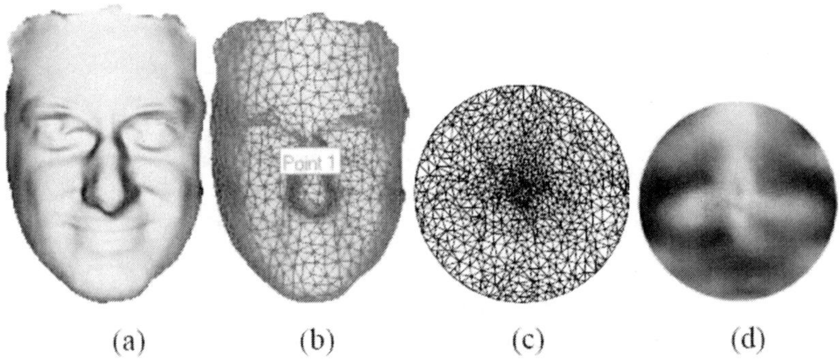

Figure 7. Local shape description at point 1 for the face model A: (a) Original range image; (b) Triangle meshes in 3D space; (c) Planar meshes after Harmonic Mapping; (d) LSD with mean curvature texture (Li et al., 2009)

4. ACTIVE APPEARANCE MODEL APPROACH

Many researchers have used the active appearance model (AAM) (Cootes et al., 2001) in modelling 3D face images. The AAM is a generative and parametric model that allows representation of a variety of shapes and appearances of human faces. It uses the basis vectors that are obtained by applying principal component analysis (PCA) to the input images and tries to find the maximum amount of variance. Although AAM is simple and fast, fitting it to an input image is not an easy task because it requires nonlinear optimization that finds a set of suitable parameters simultaneously, and its computation is basically conducted in an iterative manner. Usually, the fitting is performed by a variety of standard nonlinear optimization methods.

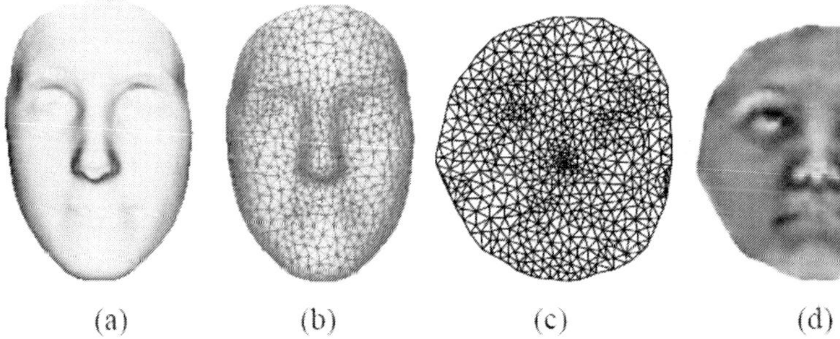

Figure 8. Global shape description for the face model B: (a) Original range image; (b) Triangle meshes in 3D space; (c) Planar meshes after ABF++; (d) GSD with mean curvature texture (Li et al., 2009)

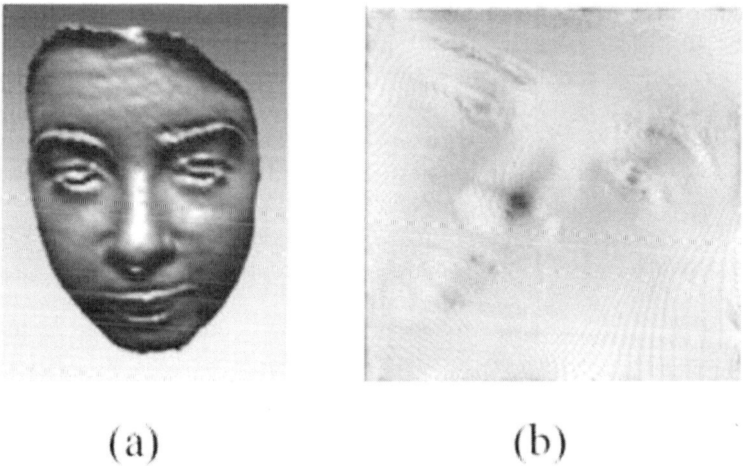

Figure 9. Constrained conformal mapping result (a) original 3D model (b) the mapping result of (a) (Guo et al., 2010)

(Abboud et al., 2004) proposed the facial expression synthesis and recognition system by face model with AAM. After extracting appearance parameters of AAM for recognition, they recognized facial expression in Euclidian and Mahalanobis space of these parameters. Also, (Abboud & Davoine, 2004) proposed a bilinear factorization expression classifier for the recognition and compared it to linear discriminant analysis (LDA). Their results showed that the bilinear factorization is useful when only a few number of training samples are available. (Ishikawa et al., 2004) used AAM for tracking around the eye region and recognized the direction of gaze.

(Matthews et al., 2004) suggested that the performance of an AAM built with single-person data is better than that of AAM built with multiple person data for the pose and illumination problems. (Xiao et al., 2004) employed 3D shapes in the AAM in order to solve the pose problem and used a nonrigid structure-from-motion algorithm for computing this 3D shape from 2D images. The 3D shape provides the constraints on the 2D shape, which can be more deformable, and these constraints make fittingmore reliable. (Hu et al., 2004) proposed another extension of a 2D + 3D AAM fitting algorithm, called the multiview AAM fitting algorithm. It fits a single 2D + 3D AAM to multiple view images obtained simultaneously from multiple affine cameras. (Mittrapiyanuruk et al., 2004) proposed the use of stereo vision to construct a 3D shape and estimate the 3D pose of a rigid object using AAM. (Cootes et al., 2002)proposed using several face models to fit an input image. They estimated the pose of an input face image by a regression technique and then fitted the input face image to the face model closest to the estimated pose. However, their approach requires pose estimation, which is another difficult problem, since the pose estimation might cause an incorrect result when the appearance of the test face image is slightly different from the training images due to different lighting conditions or different facial expressions. (Sung & Kim, 2008) proposed an extension of the 2D + 3D AAM to a viewbased approach for pose-robust face tracking and facial expressions. They used the PCA with missing data (PCAMD) technique to obtain the 2-D and 3-D shape basis vectors since some face models have missing data. Then, they developed an appropriate model selection for the input face image.Thismodel selection method uses the pose angle that is estimated from the 2D + 3D AAM directly.

(Park et al., 2010) proposed a method for aging modelling in the 3D domain. Facial aging is a complex process that affects both the shape and texture (e.g., skin tone or wrinkles) of a face. This aging process also appears in different manifestations in different age groups. While facial aging is mostly represented by facial growth in younger age groups (e.g., ≤ 18 years old), it is mostly represented by relatively large texture changes and minor shape changes (e.g., due to change of weight or stiffness of skin) in older age groups (e.g., >18). Therefore, an age correction scheme needs to be able to compensate for both types of aging processes. (Park et al., 2010) have shown how to build a 3D aging model given a 2D face aging database. Further, they have compared three different modelling methods, namely, shape modelling only, separate shape and texture modelling, and combined shape and texture modelling (e.g., applying second level PCA to remove the correlation between shape and texture after concatenating the two types of feature vectors).

5. FILTERING BASED APPROACH

(Yang et al., 2008) applied the canonical correlation analysis (CCA) to learn the mapping between the 2D face image and 3D face data. The proposed method consists of two phases. In the learning phase, given the 2D-3D face data pairs of the subjects for training, PCA is first applied on both 2D face image and 3D face data to avoid the curse of dimensionality and reduce noise. Then the CCA regression is performed between the features of 2D-3D in the previous PCA subspaces. In the recognition phase, given an input 2D face image as a probe, the correlation between the probe and the gallery is computed as matching score using the learnt regression. Furthermore, to simplify the mapping between 2D face image and 3D face data, a patch based strategy is proposed to boost the accuracy of matching. (Huang et al., 2010) presented an asymmetric 3D-2D face recognition method, that uses textured 3D face image for enrolment while performs automatic identification using only 2D facial images. The goal is to limit the use of 3D data to where it really helps to improve face recognition accuracy. The proposed method contains two separate matching steps: Sparse Representation Classifier (SRC) which is applied to 2D-2D matching, and CCA which is exploited to learn the mapping between range local binary pattern (LBP) faces (3D) and texture LBP faces (2D). Both matching scores are combined for the final decision.

(Günlü & Bilge, 2010) divided 3D faces into smaller voxel regions and applied 3D transformation to extract features from these voxel regions, as shown in Fig. 10. The number of features selected from each voxel region is not constant and depends on their discrimination.

(Dahm & Gao, 2010) presented a novel face recognition approach that implements cross-dimensional comparison to solve the issue of pose invariance. The approach implements a Gabor representation during comparison to allow for variations in texture, illumination, expression and pose. Kernel scaling is used to reduce comparison time during the branching search, which determines the facial pose of input images. This approach creates 2D rendered views of the 3D model from different angles, which are then compared against the 2D probe. Each rendered view is created by deforming the 3D model's texture with the 3D shape information, as shown in Fig. 11.

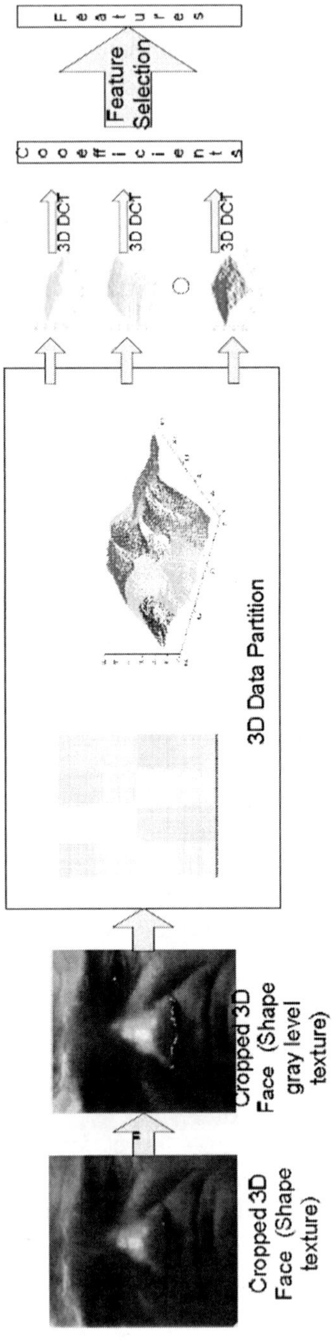

Figure 10. Proposed method by (Günlü & Bilge, 2010)

(Wang et al., 2010) proposed another scheme for 3D face recognition that passes through different stages. They used iterative closet point to align all 3D face images with the first person. Then a region defined by a sphere of radius 100 mm centred at the nose tip was cropped to construct the depth image. The Gabor filter was used to capture the useful local structure of the depth images.

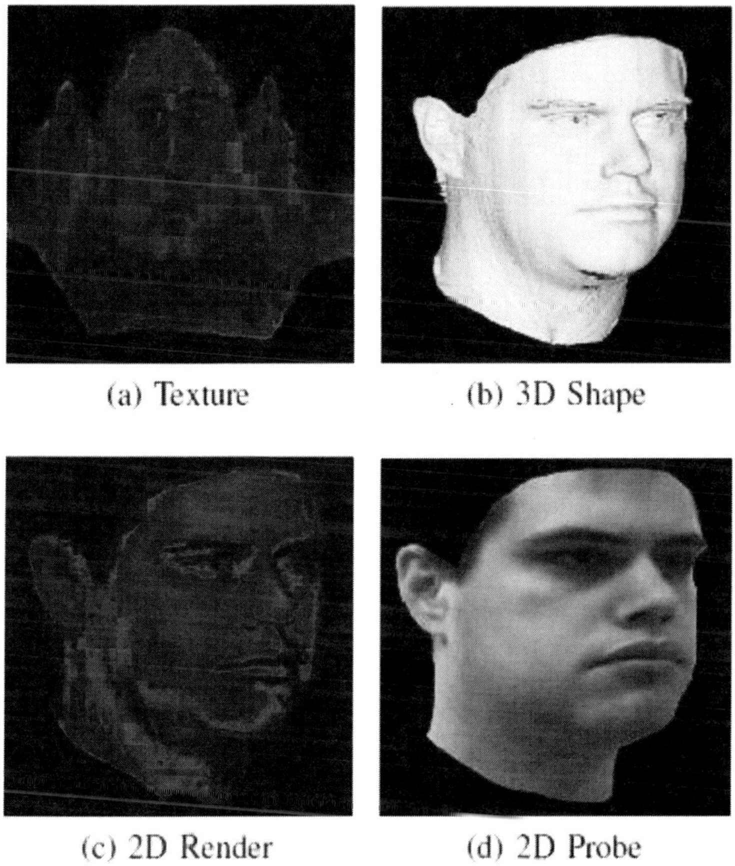

Figure 11. Comparison of 3D and 2D representation. Contrast and Brightness have been increased on texture and render for viewing (Dahm & Gao, 2010)

Another approach that deals with 3D face recognition was presented by (Cook et al., 2007), where they used multi-scale techniques to partition the information contained in the frequency domain prior to dimensionality reduction. In this manner, it is possible to increase the information available for classification and, hence, increase the discriminative performance of both Eigenfaces and Fisherfaces techniques, which were used for dimensionality reduction. They have used the Gabor filters as a partitioning scheme, and compared their results against the discrete cosine transform and the discrete wavelet transform.

6. STATISTICAL APPROACH

(Rama & Tarrés, 2005) have presented Partial Principal Component Analysis (P^2CA) for 3D face recognition. The main advantage in comparison with the model-based approaches is its low computational complexity since P^2CA does not require any fitting process. However, one of the main problems of their work is the enrolment of new persons in the database (gallery set) since a total of five different images are needed for getting the 180º texture map. Recently, they presented a work that automatically creates 180º texture maps from only two images (frontal and profile views) (Rama & Tarrés, 2007). Nevertheless, this work has also another constraint; it needs a normalization (registration) process for both eyes where they should be perfectly aligned at a fixed distance. Thus, errors in the registration of the profile view lead to noisy areas of the reconstructed 180º images (Fig. 12).

(Gupta et al., 2007) presented a systematic procedure for selecting facial fiducial points associated with diverse structural characteristics of a human face. They have identified such characteristics from the existing literature on anthropometric facial proportions. Also, they have presented effective face recognition algorithms, which employ Euclidean/geodesic distances between these anthropometric fiducial points as features along with linear discriminant analysis (LDA) classifiers. They have demonstrated how the choice of facial fiducial points critically affects the performance of 3D face recognition algorithms that employ distances between them as features.

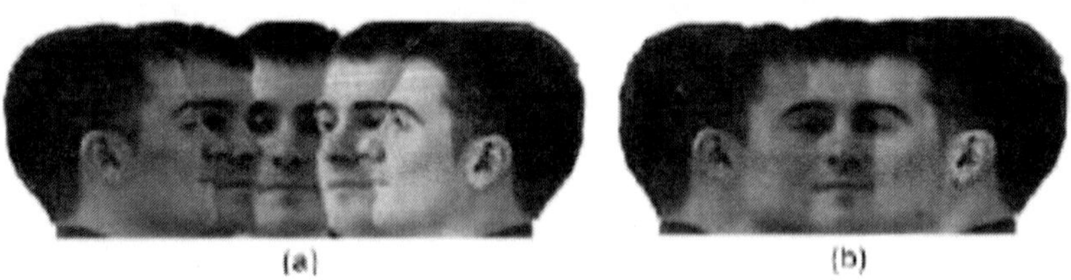

Figure 12. a) Set of images used for the creation of the training data; (b) Example of a 180º texture training image (Rama & Tarrés, 2007)

Anthropometry is the branch of science that deals with the quantitative description of physical characteristics of the human body. Anthropometric cranio-facial proportions are ratios of pairs of straight-line and/or along-the-surface distances between specific cranial and facial fiducial points (Fig. 13).

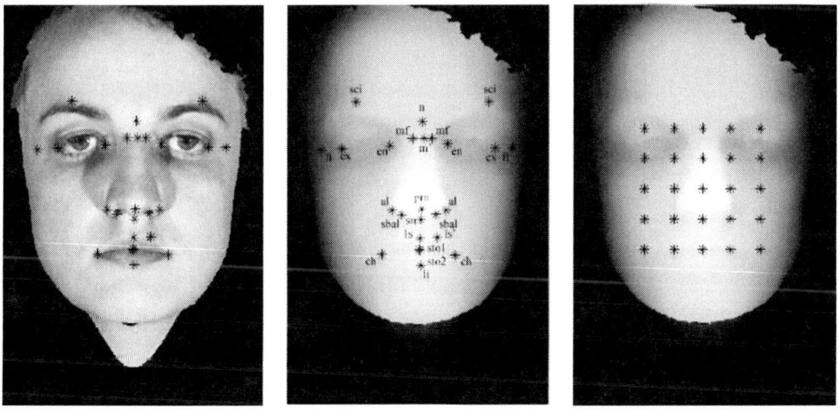

(a) (b) (c)

Figure 13. The figure depicts (a) 25 anthropometric fiducial points on a texture image; (b) 25 anthropometric fiducial points on a range image; (c) 25 arbitrary equally spaced points overlaid on the main facial features (Gupta et al., 2007)

(Ming et al., 2010) proposed algorithm for 3D-based face recognition by representing the facial surface, by what is called a Bending Invariant (BI), invariant to isometric deformations resulting from expressions and postures. In order to encode relationships in neighbouring mesh nodes, Gaussian-Hermite moments are used for the obtained geometric invariant, which provide rich representation, due to their mathematical orthogonality and effectiveness in characterizing local details of the signal. Then, the signature images are decomposed into their principle components based on Spectral Regression Kernel Discriminate Analysis (SRKDA) resulting in a huge time saving.

7. LOCAL BINARY PATTERNS

In (Zhou et al., 2010), Local Binary Patterns (LBP) method was used to represent 3D face images. The Local Binary Pattern (LBP) method describes the local texture pattern with a binary code. It is built by thresholding a neighbourhood *P* with radius *R* (typically denoting the 8 surrounding pixels) by the gray value *g* of its centre *c*. Also, (Ming et al., 2010) proposed a framework for 3D face recognition that is based on the 3D Local Binary Patterns (3D LBP). In the feature extraction stage, 3D LBP is adopted to describe the intrinsic geometric information, negating the effect of expression variations effectively. 3D LBP encodes relationships in neighbouring mesh nodes and own more potential power to describe the structure of faces than individual points. In learning stage, Spectral Regression is adopted to learn principle components from each 3D facial image. With dimensional reduction based on Spectral Regression, more useful and significant features can be produced for a face, resulting in a huge saving in computational cost. Finally, face recognition is achieved using Nearest Neighbour Classifiers.

8. OTHER 3D FACE RECOGNITION APPROACHES

In order to enhance robustness to expression variations, a procedure for 3D face recognition based on the depth image and Speeded-Up Robust Features (SURF) Operator was proposed by (Yunqi et al., 2010). First, they have applied the Fisher Linear Discriminant (FLD) method on the depth image to perform coarse recognition to catch the highly ranked 3D faces. On the basis of this step, they extracted the SURF features of the 2D gray images that are corresponding only to those highly ranked 3D faces, to carry out the refined recognition. SURF algorithm was first proposed by (Bay et al., 2008). At present, SURF has been applied to image registration, camera calibration and object recognition. Furthermore, (Kim & Dahyot, 2008) presented another approach for 3D face recognition usingSVM and SURF Descriptor.

On the other hand, (Wang et al., 2009) used a spherical harmonic representation with the morphable model for 2D face recognition. The method uses a 2D image to build a 3D model for the gallery, based on a 3D statistical morphable model. Also, (Biswas et al., 2009) proposed a method for albedo estimation for face recognition using two-dimensional images. However, they assumed that the image did not contain shadows. (Zhou et al., 2008) used nearest-subspace patch matching to warp near frontal face images to frontal and project this face image into a pre-trained low-dimensional illumination subspace. Their method requires training of patches in many different illumination conditions.

9. 3D FACE FITTING

A 3D Morphable Model (3DMM) consists of a parameterized generative 3D shape, and a parameterized albedo model together with an associated probability density on the model coefficients. Together with projection and illumination parameters, a rendering of the face can be generated. Given a face image, one can also solve the inverse problem of finding the coefficients which most likely generated the image. Identification and manipulation tasks in coefficient space are trivial, because the generating factors (light, pose, camera, and identity) have been separated. Solving this inverse problem is termed "model fitting", and was introduced for faces by (Blanz & Vetter, 1999). A similar method has also been applied to stereo data (Amberg et al., 2007) and 3D scans (Amberg et al., 2008).

A 3D deformation modelling scheme was proposed by (Lu & Jain, 2008) to handle the expression variations. They proposed a facial surface modelling and matching scheme to match 2.5D facial scans in the presence of both nonrigid deformations and pose changes (multiview) to a stored 3D face model with neutral expression.

They collected data for learning 3D facial deformations from only a small group of subjects, called the control group. Each subject in the control group provides a scan with neutral expression and several scans with nonneutral expressions. The deformations (between neutral scan and nonneutral scans) learned from the control group are transferred to and synthesized for all the 3D neutral face models in the gallery, yielding deformed templates with synthesized expressions (Fig. 14). For each subject in the gallery, deformable models are built based on the deformed templates. In order to learn deformation from the control group, a set of fiducial landmarks is needed. Besides the fiducial facial landmarks such as eye and mouth corners, landmarks in the facial area with little texture, for example, cheeks are extracted in order to model the 3D surface movement due to expression changes. A hierarchical geodesic-based resampling scheme constrained by fiducial landmarks is designed to derive a new landmark-based surface representation for establishing correspondence across expressions and subjects.

(Wang et al., 2009) proposed an improved algorithm aiming at recognizing faces of different poses when each face class has only one frontal training sample. For each sample, a 3D face is constructed by using 3DMM. The shape and texture parameters of 3DMM are recovered by fitting the model to the 2D face sample which is a non-linear optimization problem. The virtual faces of different views are generated from the 3DMM to assist face recognition. They have located 88 sparse points from the 2D face sample by automatic face fitting and used their correspondence in the 3D face as shape constraint (Fig. 15).

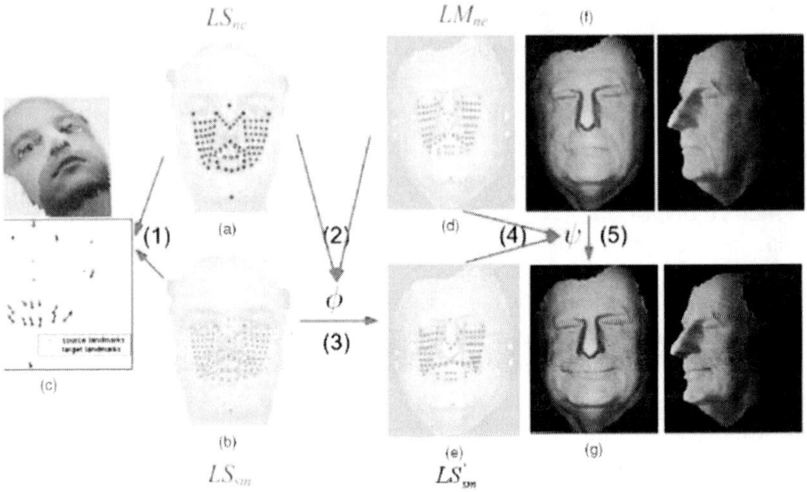

Figure 14. Deformation transfer and synthesis (Lu & Jain, 2008)

(Daniyal et al., 2009) proposed a compact face signature for 3D face recognition that is extracted without prior knowledge of scale, pose, orientation or texture. The automatic extraction of the face signature is based on fitting a trainedPoint Distribution Model (PDM) (Nair & Cavallaro, 2007). First, a facial representation based on testing extensive sets of manually selected landmarks is chosen. Next, a PDM is trained to identify the selected set of landmarks (Fig. 16). The recognition algorithm represents the geometry of the face by a set of Inter-Landmark Distances (ILDs) between the selected landmarks. These distances are then compressed using PCA and projected onto the classification space using LDA. The classification of a probe face is finally achieved by projecting the probe onto the LDA-subspace and using the nearest mean classifier.

(Paysan et al., 2009) proposed a generative 3D shape and texture model, the Basel Face Model (BFM). The model construction passes through four steps: 3D face scanning, Registration, Texture Extraction and Inpainting, and Model. The model is based on parameterizing the faces using triangular meshes. A face is then represented by two dimensional vectors: shape and texture, constructing two independent Linear Models. Finally, a Gaussian distribution is fit to the data using PCA (Fig. 17).

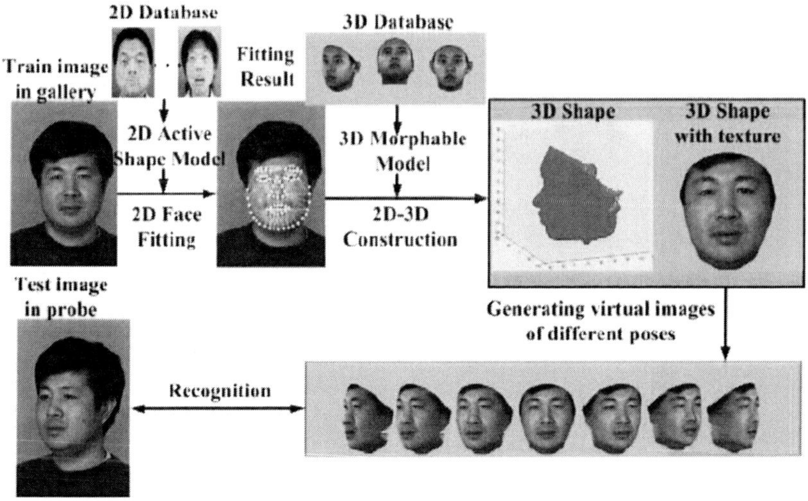

Figure 15. Algorithm overview (Wang et al., 2009)

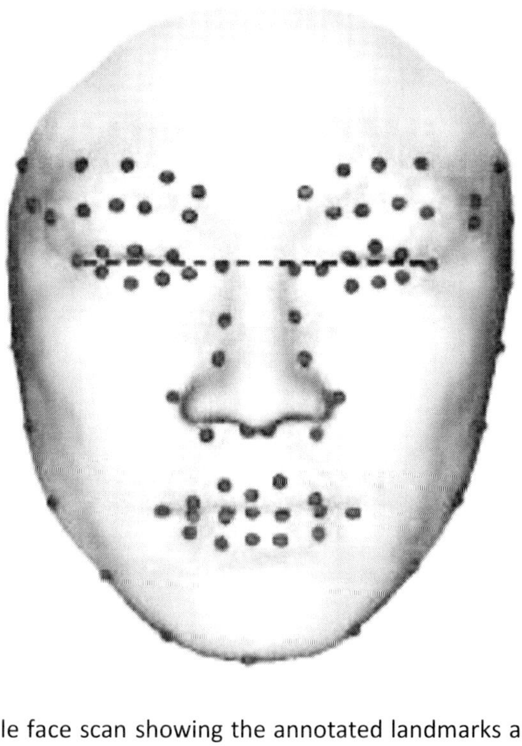

Figure 16. Sample face scan showing the annotated landmarks and the scaling distance *dS* (dotted line) used in (Daniyal et al., 2009)

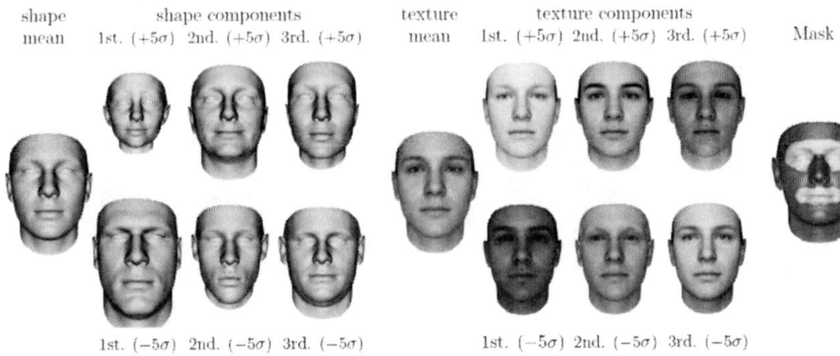

Figure 17. The mean together with the first three principle components of the shape (left) and texture (right) PCA model (Paysan et al., 2009)

(Toderici et al., 2010) proposed a face recognition method which utilizes 3D face data for enrolment, while it requires only 2D data for authentication. During enrolment, 2D+3D data (2D texture plus 3D shape) is used to build subject-specific annotated 3D models. First, an Annotated Face Model (AFM) is fitted to the raw 2D+3D data using a subdivision based deformable framework. Then, a geometry image representation is extracted using the UV parameterization of the model. In the authentication phase, a single 2D image is used as the input to map the subject-specific 3D AFM. After that, an Analytical Skin Reflectance Model (ASRM) is applied to the gallery AFM in order to transfer the lighting from the probe to the texture in the gallery.

10. FACE RECOGNITION IN VIDEO

Face recognition in video has gained wide attention as a covert method for surveillance to enhance security in a variety of application domains (e.g., airports). A video contains temporal information (e.g., movements of facial features) as well as multiple instances of a face, so it is expected to lead to a better face recognition performance compared to still face images. However, faces appearing in a video have substantial variations in pose and lighting. These pose and lighting variations can be effectively modelled using 3D face models (Yin et al., 2006). Given the trajectories of facial feature movement, face recognition is performed based on the similarities of the trajectories. The trajectories can also be captured as nonlinear manifolds and the distance between clusters of faces in the feature space establishes the identity associated with the face. Production of 3D faces from video can be performed using morphable models, stereography, or structure from motion (SFM).

(Park et al., 2005) proposed a face recognition system that identifies faces in a video using 3D face model. Ten video files were recorded for ten subjects under four different lighting conditions at various poses with yaw and pitch motion. Recognition using multiple images and temporal cue was explored and majority voting and score sum were used to fuse the recognition result from multiple frames. To use temporal cues for the recognition, a LDA based classifier was used. After the face pose in a video was estimated, frames of different poses under specific lighting condition and specific order were extracted to form a probe sequence.

(Von Duhn et al., 2007) designed a 3D face analyzer using regular CCTV videos. They used a three view tracking approach to build 3D face models over time. The proposed system detects, tracks and estimates the facial features. For the tracking, an Active Appearance Model approach is adapted to decrease the amount of manual work that must be done. After the tracking stage, a generic model is adapted to the different views of the face using a face adaptation algorithm, which includes two steps: feature point adaptation and non-feature point interpolation. Finally, the multiple views of models are combined to create an individualized face model. To track the facial motion under three different views, i.e., front view, side view, and angle view, predefined fiducial points are used.

Also, (Roy-Chowdhury & Xu, 2006) estimated the pose and lighting of face images contained in video frames and compared them against synthetic 3D face models exhibiting similar pose and lighting. However, the 3D face models were registered manually with the face image in the video. (Lee et al., 2003) proposed an appearance manifold based approach where each database or gallery image was matched against the appearance manifold obtained from the video. The manifolds were obtained from each sequence of pose variations. (Zhou et al., 2003) proposed to obtain statistical models from video using low level features (e.g., by PCA) contained in sample images. The matching was performed between a single frame and the video or between two video streams using the statistical models.

(Park et al., 2007) explored the adaptive use of multiple face matchers in order to enhance the performance of face recognition in video. To extract the dynamic information in video, the facial poses in various frames are explicitly estimated using Active Appearance Model and a Factorization based 3D face reconstruction technique. The motion blur is estimated using Discrete Cosine Transformation (DCT). The performance of the proposed system could be improved by dynamically fusing the matching results from multiple frames and multiple matchers.

Further, (Wang et al., 2004) have successfully developed a hierarchical framework for tracking high density 3D facial expression sequences captured from a structure-lighting imaging system. The work in (Chang et al., 2005), utilized six 3D model sequences for facial analysis and editing. The work was mainly for facial expression analysis. (Papatheodorou & Rueckert, 2004) evaluated a so-called 4D face recognition approach, which was, however, just the 3D static data plus texture, no temporal information was explored. (Li et al., 2003) reported a model fitting approach to generate facial identity surfaces through video sequences. The application of this model to face recognition relies on the quality of the tracked low resolution face model.

(Sun & Yin, 2008) proposed to use a Spatio-Temporal Hidden Markov Model (HMM) which incorporates 3D surface feature characterization to learn the spatial and temporal information of faces. They have created a face database including 606 3D model sequences with six prototypic expressions. To evaluate the usability of such data for face recognition, they applied a generic model to track the range model sequences and establish the correspondence of range model frames over time. After the tracking model labelling and LDA transformation, they trained two HMM models (S-HMM and T-HMM) for each subject to learn the spatial and temporal information of the 3D model sequence. The query sequence was classified based on the results of the two HMMs.

(Medioni et al., 2007) utilized synthetic stereo to model faces in a 3048 x 4560 video stream. By tracking the pose and location of the face, a synthetic stereo rig based upon the different poses between two frames is initialized. Multiple point clouds from different stereo pairs are created and integrated into a single model. (Russ et al., 2006) utilized a 3D PCA based approach for face recognition. The approach determines a correspondence that utilizes a reference face aligned via ICP to determine a unique vector input into PCA. The coefficients from PCA are used to determine the identity as in 2D PCA face recognition. (Kakadiaris et al., 2006) converted the 3D model into a depth map image for wavelet analysis. This approach performs well and does not utilize ICP as the basis for each match score computation, but does for the depth map production.

Moreover, (Boehnen & Flynn, 2008) presented an approach to combine multiple noisy low density 3D face models obtained from uncalibrated video into a higher resolution 3D model using SFM method. SFM is a method for producing 3D models from a calibrated or uncalibrated video stream utilizing equipment that is inexpensive and widely available. The approach first generates ten 3D face models (containing a few hundred vertices each) of each subject using 136 frames of video data in which the subject face moves in a range of approximately 15 degrees from frontal. By aligning, resampling, and merging these models, a new 3D face model containing over 50,000 points is produced. An ICP face matcher employing the entire face achieved a 75% rank one recognition rate.

Using a data set of varying facial expressions and lighting conditions, (Bowyer et al., 2006) reported an improvement in rank one recognition rate from 96.11% with two frames per subject to 100% with four frames per subject. In another study, (Thomas et al., 2007) observed that the recognition rate generally increases as the number of frames per subject increases, regardless of the type of camera being used. They also found that the optimal number of frames per subject is between 12 and 18, given the particular data sets used.

(Canavan et al., 2007) discussed that the 3D geometry of a rotating face can be embedded in the continuous intensity changes of an image stream, and therefore the recognition algorithm does not require an explicit 3D face model. Further, multiple video frames that capture the face at different pose angles can be combined to provide a more reliable and comprehensive 3D representation of the face than any single view image. Also, they have discussed that a video sequence of a face with different poses might help alleviate the adverse effect of lighting changes on recognition accuracy. For instance, a light source can cast shadows on a face, but at the same time, it also reveals the 3D curvatures of the face by creating sharp intensity contrasts (such as silhouette).

(Dornaika & Davoine, 2006) introduced a view- and texture-independent approach that exploits the temporal facial action parameters estimated by an appearance-based 3D face tracker. The facial expression recognition is carried out using learned dynamical models based on auto-regressive processes. These learned models can also be utilized for the synthesis and prediction tasks. In their study, they used the 3D face model *Candide* (Ahlberg, 2001). This 3D deformable wireframe model is given by the 3D coordinates of the vertices P_i, $i = 1,\ldots, n$ where n is the number of vertices. Thus, the shape up to a global scale can be fully described by the $3n$-vector g, the concatenation of the 3D coordinates of all vertices P_i. The vector g can be written as:

$$g = \overline{g} + S\tau_s + A\tau_a \qquad (1)$$

where $\overline{g}$ is the standard shape of the model, and the columns of S and A are the shape and action units, respectively. Thus, the term $S\tau_s$ accounts for shape variability (inter-person variability) while the term $A\tau_a$ accounts for the facial action (intra-person variability).

11. CONCLUSION

In this chapter, we have presented a study on the most recent advancements in 3D face recognition field. Despite the huge developments made in this field, there are still some problems and issues which need to be resolved.

Due to the computational complexity, fussy pre-treatment, and expensive equipment, 3D technology is still not used widely in practical applications. To acquire an accurate 3D face data, some very costly equipment must be used, such as 3D laser scan or stereo camera system. Also, they are still not as stable and efficient as 2D cameras, and for some cases like the stereo camera system, calibration is needed before use. Moreover, they take a longer time to acquire (or reconstruct) when compared to the 2D camera. Further, 3D data require much more storage space. Other challenges include feature points allocation (this is still a debatable topic) that is also sensitive to the quality of data. Sampling density of the facial surface and accuracy of the depth, are among the issues that require more investigations. Furthermore, no standard testing protocol is available to compare between different 3D face recognition systems.

On the other hand, in video-based face recognition, experiments have shown that multi-frame fusion is an effective method to improve the recognition rate. The performance gain is probably related to the use of 3D face geometry embedded in video sequences. However, it is not clear how the inter-frame variation has contributed to the observed performance increase. Will the multi-frame fusion work for videos of strong shadows? How many frames are necessary for maximizing the recognition rate without incurring a heavy computational cost? To address these issues, more exploration is needed from the research community.

ACKNOWLEDGEMENTS

The author would like to acknowledge and thank Kuwait Foundation for the Advancement of Sciences (KFAS) for financially supporting this work.

REFERENCES

1. B. Abboud, F. Davoine, 2004 Appearance factorization for facial expression recognition and synthesis, Proceedings of Int. Conf. Pattern Recog., 163 166 , 2004

2. B. Abboud, F. Davoine, M. Dang, 2004 Facial expression recognition and synthesis based on an appearance model. Signal Process. Image Commun., 19 8 2004, 723 740

3. J. Ahlberg, 2001 CANDIDE-3 - an updated parametrized face, Tech. Rep. LiTH-ISY-R-2326, Dept of Electrical Engineering, Link°oping University, Sweden

4. B. Amberg, R. Knothe, T. Vetter, 2008 Expression invariant 3D face recognition with a morphable model, Proceedings of FG'08, 2008

5. B. Amberg, S. Romdhani, A. Fitzgibbon, A. Blake, T. Vetter, 2007 Accurate surface extraction using model based stereo, Proceedings ofICCV, 2007

6. H. Bay, T. Tuytelaars, I. Van Gool, 2008 Speeded-Up Robust Features (SURF). Computer Vision and Image Understanding, 2008 110346 359

7. S. Biswas, G. Aggarwal, R. Chellappa, 2009 Robust estimation of albedo for illumination-invariant matching and shape recovery. IEEE Transactions on Pattern Analysis and Machine Intelligence, 31 2009, 884 899

8. V. Blanz, T. Vetter, 1999 A morphable model for the synthesis of 3D faces, Proceedings of SIGGRAPH, 1999

9. C. Boehnen, P. J. Flynn, 2008 Increased Resolution 3D Face Modeling and Recognition From Multiple Low Resolution Structure From Motion Models, Proceedings of IEEE, 2008

10. K. W. Bowyer, K. Chang, P. J. Flynn, X. Chen, 2006 Face recognition using 2-D, 3-D and Infrared: Is multimodal better than multisample?, Proceedings of IEEE, 94 11 2000 2012 , 2006

11. A. M. Bronstein, M. M. Bronstein, R. Kimmel, 2007 Expression-invariant representations of faces. IEEE Trans. on PAMI, 2007, 1042 1053

12. S. J. Canavan, M. P. Kozak, Y. Zhang, J. R. Sullins, M. A Shreve, D. B. Goldgof, 2007 Face Recognition by Multi Frame Fusion of Rotating Heads in Videos, Proceedings of IEEE, 2007

13. Y. Chang, M. Vieira, M. Turk, L. Velho, 2005 Automatic 3d facial expression analysis in videos, Proceedings of ICCV Workshop on Analysis and Modeling of Faces and Gestures, 2005

14. J. Cook, V. Chandran, C. Fookes, 2006 3D face recognition using log-gabor templates, Proceedings of the 17th British Machine Vision Conference, 2006

15. J. Cook, V. Chandran, S. Sridharan, 2007 Multiscale Representation For 3-D Face Recognition. IEEE Transactions on Information Forensics and Security, 2 3 September 2007

16. T. F. Cootes, G. J. Edwards, C. J. Taylor, 2001 Active appearance models, IEEE Trans. Pattern Anal. Mach. Intell., 23 6 Jun. 2001, 681 685

17. T. F. Cootes, G. V. Wheeler, K. N. Walker, C. J. Taylor, 2002 Viewbased active appearance models. Image Vis. Comput., 20 9 Aug. 2002, 657 664

18. N. Dahm, Y. Gao, 2010 A Novel Pose Invariant Face Recognition Approach Using A 2D-3D Searching Strategy, Proceedings of Int'l Conf. on Pattern Recognition, 2010

19. F. Daniyal, P. Nair, A. Cavallaro, 2009 Compact signatures for 3D face recognition under varying expressions, Proceedings of Advanced Video and Signal Based Surveillance, 2009

20. F. Dornaika, F. Davoine, 2006 Facial Expression Recognition using Auto-regressive Models , Proceedings of18th International Conf. on Pattern Recognition, 2006

21. H. Drira, B. B. Amor, M. Daoudi, A. Srivastava, 2009 A riemannian analysis of 3d nose shapes for partial human biometrics, Proceedings ofICCV, 1 1 1 8 , 2009

22. S. V. Duhn, L. Yin, M. J. Ko, T. Hung, 2007 Multiple-View Face Tracking For Modeling and Analysis Based On Non-Cooperative Video Imagery, Proceedings of IEEE, 2007

23. E. Elyan, H. Ugail, 2009 Automatic 3D Face Recognition Using Fourier Descriptors, International Conference on CyberWorlds, 2009

24. T. Faltemier, K. W. Bowyer, P. J. Flynn, 2008 A region ensemble for 3D face recognition. IEEE Trans. on Information Forensics and Security, 3 1 2008, 62 73

25. Z. Guo, Y. Zhang, Z. Lin, D. Feng, 2009 A Method Based on Geometric Invariant Feature for 3D Face Recognition, Proceedings of Fifth International Conference on Image and Graphics, 2009

26. Z. Guo, Y. Zhang, Y. Xia, Z. Lin, D. Feng, 2010 3D Face Representation And Recognition By Intrinsic Shape Description Maps, Proceedings of ICASSP, 2010

27. S. Gupta, J. K. Aggarwal, M. K. Markey, A. C. Bovik, 2007 3D Face Recognition Founded on the Structural Diversity of Human Faces, Proceedings of IEEE, 2007

28. G. Günlü, H. S. Bilge, 2010 3D Face Decomposition and Region Selection against Expression Variations, Proceedings ofInt'l Conference on Pattern Recognition, 2010

29. J. Harguess, J. K. Aggarwal, 2009 A Case for the Average-Half-Face in 2D and 3D for Face Recognition, Proceedings of IEEE, 2009

30. C. Hu, J. Xiao, I. Matthews, S. Baker, J. Cohn, T. Kanade, 2004 Fitting a single active appearance model simultaneously to multiple images, Proceedings of Brit. Mach. Vis. Conf., 2004

31. D. Huang, M. Ardabilian, Y. Wang, L. Chen, 2010 Automatic Asymmetric 3D-2D Face Recognition, International Conference on Pattern Recognition, 2010

32. T. Ishikawa, S. Baker, I. Matthews, T. Kanade, 2004 Passive driver gaze tracking with active appearance models, Robot. Inst., Carnegie Mellon Univ., Pittsburgh, PA, Tech. Rep. CMU-RI-TR-04 08 , Feb. 2004

33. I. A. Kakadiaris, G. Passalis, G. Toderici, N. Murtuza, Y. Lu, N. Karampatziakis, T. Theoharis, 2007 Three-dimensional face recognition in the presence of facial expressions: An annotated deformable model approach. IEEE Trans. on PAMI, 29 4 2007, 640 649

34. I. Kakadiaris, G. Passalis, G. Toderici, N. Murtuza, T. Theoharis, 2006 3D face recognition, Proceedings of the British Machine Vision Conference, 200 208 , 2006

35. D. Kim, R. Dahyot, 2008 Face components detection using SURF descriptors and SVMs. Proceedings of Int'l Machine Vision and Image Processing Conf., 51 56 , 2008

36. K. C. Lee, J. Ho, M. H. Yang, D. Kriegman, 2003 Video- Based Face Recognition using probabilistic appearance manifolds, Proceedings of Intl. Conf. on Computer Vision and Pattern Recognition, 2003

37. W. Li, Z. Yin, J. Wu, Y. Xiong, 2009 3D Face Recognition Based on Local/Global Shape Description, Proceedings of Int'l Conf. on Information Technology and Computer Science, 2009

38. Y. Li, S. Gong, H. Liddell, 2003 Constructing facial identity surfaces for recognition. Int'l Journal of Comp. Vision, 53 1 2003, 71 92

39. X. Lu, A. K. Jain, 2008 Deformation Modeling For Robust 3D Face Matching. IEEE Transactions On Pattern Analysis And Machine Intelligence, 30 8 August 2008, 1346 1356

40. I. Matthews, S. Baker, R. Gross, 2004 Generic vs. person specific active appearance models, Proceedings of Brit. Mach. Vis. Conf., 1080 1093 , Sep. 2004

41. T. Maurer, D. Guigonis, I. Maslov, B. Pesenti, A. Tsaregorodtsev, D. West, G. Medioni, I. Geometrix, 2005 Performance of Geometrix Active ID 3D Face Recognition Engine on the FRGC Data, Proceedings ofIEEE CVPR, 154 154 , 2005

42. C. Mc Cool, V. Chandran, S. Sridharan, C. Fookes, 2008 3D face verification using a free-parts approach. Pattern Recogn. Lett., 29 9 2008

43. G. Medioni, D. Fidaleo, J. Choi, L. Zhang, C. Kuo, H. , K. Kim, 2007 Recognition of Non-Cooperative Individuals at a Distance with 3D Face Modeling, Proceedings of IEEE Workshop on Automatic Identification Advanced Technologies, 2007

44. Y. Ming, Q. Ruan, R. Ni, 2010 Learning Effective Features For 3D Face Recognition, Proceedings of 17th International Conference On Image Processing, September 26-29, 2010

45. Y. Ming, Q. Ruan, X. Wang, M. Mu, 2010 Robust 3D Face Recognition using Learn Correlative Features, Proceedings ofICSP, 2010

46. P. Mittrapiyanuruk, G. N. De Souza, A. C. Kak, 2004 Calculating the 3D-pose of rigid objects using active appearance models, Proceedings of IEEE Int. Conf. Robot. Autom., 5 5147 5152 , 2004

47. I. Mpiperis, S. Malassiotis, M. G. Strintzis, 2007 3-D Face Recognition With the Geodesic Polar Representation. IEEE Transactions on Information Forensics and Security, 2 3 Sept 2007

48. P. Nair, A. Cavallaro, 2007 Region segmentation and feature point extraction on 3D faces using a point distribution model, Proceedings of IEEE Intl. Conf. on Image Processing, 3 85 88 , Texas, USA, Sept, 2007

49. T. Papatheodorou, D. Rueckert, 2004 Evaluation of automatic 4d face recognition using surface and texture registration, Proceedings of the International Conference on Automatic Face and Gesture Recognition, 2004

50. U. Park, H. Chen, A. K. Jain, 2005 3D Model-Assisted Face Recognition in Video, Proceedings of the Second Canadian Conference on Computer and Robot Vision, 2005

51. U. Park, A. K. Jain, A. Ross, 2007 Face Recognition in Video: Adaptive Fusion of Multiple Matchers, Proceedings of IEEE, 2007

52. U. Park, Y. Tong, A. K. Jain, 2010 Age-Invariant Face Recognition. IEEE Trans. Pattern Analysis and Machine Intelligence, Vol. 32, No. 5, May 2010,

53. G. Passalis, I. A. Kakadiaris, T. Theoharis, 2007 Intraclass Retrieval of Nonrigid 3D Objects: Application To Face Recognition. IEEE Transactions on Pattern Analysis and Machine Intelligence, 29 2 February 2007

54. P. Paysan, R. Knothe, B. Amberg, S. Romdhani, T. Vetter, 2009 A 3D Face Model for Pose and Illumination Invariant Face Recognition, Proceedings of Advanced Video and Signal Based Surveillance, 2009

55. A. Rama, F. Tarrés, 2005 2CA A new face recognition scheme combining 2D and 3D information, Proceedings of IEEE International Conference on Image Processing, Genoa, Italy, Sept 2005

56. A. Rama, F. Tarrés, 2007 Face Recognition Using A Fast Model Synthesis From A Profile And A Frontal View, Proceedings of IEEE, ICIP, 2007

57. D. Riccio, J. L. Dugelay, 2007 Geometric invariants for 2d/3d face recognition. Pattern Recognition Letters, 28 1907 1914

58. A. Roy-Chowdhury, Y. Xu, 2006 Pose and Illumination Invariant Face Recognition Using Video Sequences. Face Biometrics for Personal Identification: Multi-Sensory Multi-Modal Systems, Springer-Verlag, 2006, 9 25

59. T. Russ, C. Boehnen, T. Peters, 2006 3D Face Recognition Using 3D Alignment for PCA, Computer Vision and Pattern Recognition, New York, 2006, 1391 1398

60. C. Samir, A. Srivastava, M. Daoudi, 2006 Three-dimensional face recognition using shapes of facial curves. IEEE Trans. Pattern Anal. Mach. Intell., 28 11 Nov. 2006, 1858 1863

61. A. Sheffer, et al. 2005 ABF++: fast and robust angle based flattening. ACM Transactions on Graphics, 24 2 Apr 2005, 311 330

62. Y. Song, W. Wang, Y. Chen, 2009 Research on 3D Face Recognition Algorithm, First International Workshop on Education Technology and Computer Science, 2009

63. Y. Sun, L. Yin, 2008 3D Spatio-Temporal Face Recognition Using Dynamic Range Model Sequences, Proceedings of IEEE, 2008

64. J. Sung, D. Kim, 2008 Pose-Robust Facial Expression Recognition Using View-Based 2d + 3d AAM. IEEE Transactions on Systems, Man, and Cybernetics-Part A: Systems and Humans, 38 4 July 2008, 852 866

65. P. Szeptycki, M. Ardabilian, L. Chen, W. Zeng, D. Gu, D. Samaras, 2010 Partial face biometry using shape decomposition on 2D conformal maps of faces, Proceedings ofInternational Conference on Pattern Recognition, 2010

66. D. Thomas, K. W. Bowyer, P. J. Flynn, 2007 Multi-frame approaches to improve face recognition, Proceedings of IEEE Workshop on Motion and Video Computing, 19 19 , Austin, Texas, 2007

67. M. M. M. Tin, M. M. Sein, 2009 Multi Triangle Based Automatic Face Recognition System By Using 3d Geometric Face Feature, International Instrumentation And Measurement Technology Conference, Singapore, May 5-7, 2009

68. G. Toderici, G. Passalis, S. Zafeiriou, G. Tzimiropoulos, M. Petrou, T. Theoharis, I. A. Kakadiaris, 2010 Bidirectional relighting for 3D-aided 2D Face Recognition, Proceedings of IEEE, 2010

69. L. Wang, L. Ding, X. Ding, C. Fang, 2009 Improved 3d Assisted Pose-Invariant Face Recognition, Proceedings ofIEEE ICASSP, 2009

70. S. Wang, Y. Wang, M. Jin, X. Gu, D. Samaras, 2006 3d surface matching and recognition using conformal geometry, Proceedings ofIEEE Conference on Computer Vision and Pattern Recognition, 2453 2460 , 2006

71. X. Wang, Q. Ruan, Y. Ming, 2010 A New Scheme for 3D Face Recognition, Proceedings of ICSP, 2010

72. Y. Wang, X. Huang, C. Lee, S. Zhang, Z. Li, D. Samaras, D. Metaxas, A. Elgammal, P. Huang, 2004 High resolution acquisition, learning and transfer of dynamic 3d facial expressions, Proceedings of EUROGRAPHICS, 2004

73. Y. Wang, L. Zhang, Z. Liu, G. Hua, Z. Wen, Z. Zhang, D. Samaras, 2009 Face relighting from a single image under arbitrary unknown lighting conditions.

IEEE Transactions on Pattern Analysis and Machine Intelligence, 31 11 Nov. 2009, 1968 1984

74. K. Wong, C. , W. Lin, Y. , Y. H. Hu, N. Boston, X. Zhang, 2007 Optimal Linear Combination Of Facial Regions For Improving Identification Performance. IEEE Transactions On Systems, Man, And Cybernetics-Part B: Cybernetics, 37 5 October 2007, pp.

75. J. Xiao, S. Baker, I. Matthews, T. Kanade, 2004 Real-time combined 2D + 3D active appearance models, Proceedings of Conf. Comp. Vis. Pattern Recog., 535 542 , 2004

76. W. Yang, D. Yi, Z. Lei, J. Sang, S. Z. Li, 2008 2D-3D Face Matching using CCA, Proceedings of IEEE, 2008

77. L. Yin, X. Wei, Y. Sun, J. Wang, M. J. Rosato, 2006 A 3D Facial Expression Database For Facial Behavior Research, Proceedings of the 7th International Conference on Automatic Face and Gesture Recognition, 2006

78. L. Yunqi, L. Haibin, J. Xutuan, 2010 3D Face Recognition by SURF Operator Based on Depth Image, Proceedings of IEEE, 2010

79. N. Zaeri, 2011 Feature extraction for 3D face recognition system, Proceedings of IEEE, 2011

80. S. Zhou, V. Krueger, R. Chellappa, 2003 Probabilistic recognition of human faces from video. Computer Vision and Image Understanding, 91 2003, 214 245

81. X. Zhou, S. A. S´anchez, A. Kuijper, 2010 3D Face Recognition with Local Binary Patterns, Proceedings of Sixth International Conference on Intelligent Information Hiding and Multimedia Signal Processing, 2010

82. Z. Zhou, A. Ganesh, J. Wright, S. F. Tsai, Y. , 2008 Nearest-subspace patch matching for face recognition under varying pose and illumination, Proceedings of 8th IEEE International Conference on Automatic Face Gesture Recognition, 1 8 , Amsterdam, The Netherlands, Sept. 2008

Chapter 7

Real-Time Video Face Recognition for Embedded Devices

Costache Gabriel[1], Mangapuram Sathish, Drimbarean Alexandru, Bigioi Petronel and Corcoran Peter

[1] Tessera, Galway, Ireland

1. INTRODUCTION

This chapter will address the challenges of real-time video face recognition systems implemented in embedded devices. Topics to be covered include: the importance and challenges of video face recognition in real life scenarios, describing a general architecture of a generic video face recognition system and a working solution suitable for recognizing faces in real-time using low complexity devices. Each component of the system will be described together with the system's performance on a database of video samples that resembles real life conditions.

2. VIDEO FACE RECOGNITION

Face recognition remains a very active topic in computer vision and receives attention from a large community of researchers in that discipline. Many reasons feed this interest; the main being the wide range of commercial, law enforcement and security applications that require authentication. The progress made in recent years on the methods and algorithms for data processing as well as the availability of new technologies makes it easier to study these algorithms and turn them into commercially viable product. Biometric based security systems are becoming more popular due to their non-invasive nature and their increasing reliability. Surveillance applications based on face recognition are gaining increasing attention after the United States' 9/11 events and with the ongoing security threats. The Face Recognition Vendor Test (FRVT) (Phillips et al., 2003) includes video face recognition testing starting with the 2002 series of tests.

Recently, face recognition technology was deployed in consumer applications such as organizing a collection of images using the faces present in the images (Picassa; Corcoran & Costache, 2005), prioritizing family members for best capturing conditions when taking pictures, or directly annotating the images as they are captured (Costache et al., 2006).

Video face recognition, compared with more traditional still face recognition, has the main advantage of using multiple instances of the same individual in sequential frames for recognition to occur. In still recognition case, the system has only one input image to make the decision if the person is or is not in the database. If the image is not suitable for recognition (due to face orientation, expression, quality or facial occlusions) the recognition result will most likely be incorrect. In the video image there are multiple frames which can be analyzed in order to have greater recognition accuracy. Even if some frames are not suitable for recognition there is a high probability that some of them will work and the decision made will have a high degree of confidence. Once a face is recognized, it remains recognized in the scene by tracking techniques.

The disadvantage in the video imaging technique is in most cases the quality and size of the input frames are inferior compared to the still images.

2.1. General Architecture Of A Vfr System

Most face recognition systems for still and video image technology follow the same classical workflow:

1 The faces have to be detected in the images.

2 The faces are normalized to the same size and usually same in-plane orientation.

3 Before or after (2), a pre-processing step tries to minimize the effect of illumination over the face.

4 Features are extracted from the facial region.

5 Test faces are compared with a database of people.

The first difference between the video and still image technology is that video scenarios can use a tracking algorithm together with a detection algorithm in order to keep track of all the faces in the video sequence. Using face tracking combined with face detection has three main advantages:

1 It allows the system to follow the faces across a wide range of variations in pose and lighting where tracking can be done easier than detection.

2 The time and memory requirements of a face tracking algorithm are lower than those of a face detection algorithm. Freed resources can be accessed once a face is detected in a frame. Tracking from that moment forward is a very important aspect when achieving real-time functionality.

3 Once a face in a particular frame is recognized with a high degree of confidence, that particular face does not need to be processed for the next frames. Only track the face and keep the association between the recognized person and the tracked face.

In the classification stage of the video imagery, a history of the recognized face offers greater accuracy than that of a still image.

Figure 1 shows a typical architecture of a video face recognition system.

Below are brief descriptions of each component together with the requirements that need to be satisfied in order to have a robust real-time face recognition system which can be integrated into an embedded device.

2.1.1. Face Detection & Tracking

The face detection and tracking component is very important in designing the recognition system. The properties of the detection algorithm (detection rate, robustness to variations, speed and memory requirements, etc.) will directly affect the properties of the overall recognition system. It is clear that undetected faces will not be recognized. Also considering the goal of real-time functionality on embedded devices where limited resources are available, spending most of that early on will reduce the application of the other blocks in the diagram in real time.

The main challenges associated with the detection and tracking algorithm are determined by the following factors:

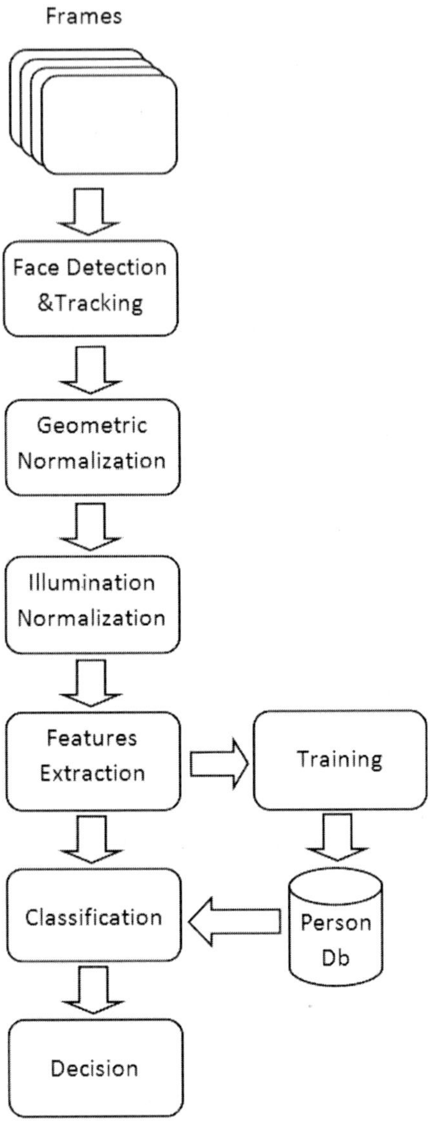

Figure 1. Architecture of a VFR system

– Face orientation (pose). The appearance of the face may differ in many ways when the orientation of the face changes from frontal to profile or extreme view angles, where face components like the eyes, the nose or the ear may be occluded. It is difficult to detect a face at these extreme angles although face tracking is achievable.

- Changes in facial appearance. Examples include beards, moustaches or glasses. Women may use make-up which can significantly alter the face color and texture. These factors together with the potential for variability in shape, size or color of the face makes face detection challenging.

- Facial expression. The appearance of the face is directly affected by the person's facial expression. Tracking has to be robust to these variations as it is likely to be encountered in normal consumer videos.

- Occlusions. Different components from the face may be occluded in the image by other objects or faces. These have to be addressed by the tracking algorithm.

- Capture conditions. Factors that are involved in capturing the image such as lighting conditions, camera characteristics or quality of the captured image may have a big influence in the detection process.

- Face size or distance to subject. For video face detection and tracking consider the capture resolution and the distance from the capture equipment to the subject. For normal working resolution (qVGA, VGA) the faces can be very small even for relatively short distances.

The detection algorithm should have a high detection rate and robustness to variations such as changes in appearance, capture conditions and face size. The tracking algorithm should improve the robustness to face orientation, expressions and occlusions.

All the above requirements are difficult to fulfil especially for real-time scenarios in embedded devices. In the last few years there has been much progress in this area and now face detection and tracking is a common feature in most consumer cameras and mobile phones.

Tessera's OptiML™ Face Tools Face Tracking and Face Recognition technologies represent a perfect example of state-of-the-art technology in this area.

Some of the relevant parameters of Tessera's Face Tools technology that affect the performance of the overall recognition system include:

- Face tracking for up to 10 faces per frame, with less than 0.1 seconds lock time

- Minimum face size: 14x14 pixels

- Real-time face tracking up to 30 frames per second
- Faces detected in a wide range of orientations including rotation-in-plane and out-of-plane

Together with the detection rate, another metric used to describe the performance of a face detection algorithm is the false positive rate which represents the number of regions falsely reported as faces by the algorithm.

The false positive rate is not as important as the detection rate because the recognition algorithm should be able to differentiate between faces and non-faces when trying to classify the false positive candidates.

2.1.2. Geometric Normalization

It is very important to detect and track the faces in all conditions and variations. When comparing local regions between faces, an image registration step must be performed so corresponding facial features are synchronised.

Simple geometric normalization usually involves bringing the faces to a standard size and rotating them in-plane in order to bring the eyes on the same horizontal line. Figure 2 shows some face samples before and after applying the geometric normalization.

More complex normalization scenarios (Corcoran et al., 2006b) can use 3D face models to rotate the face in the out-of-plane space to have identical orientation (i.e only frontal faces). This will have a higher computational requirement and could only be used when there is enough processing power.Figure 3 shows an example of the output of this complex normalization which can help recognition for large pose variations.

All other processing steps applied after geometric normalization should have the same affect on each face.

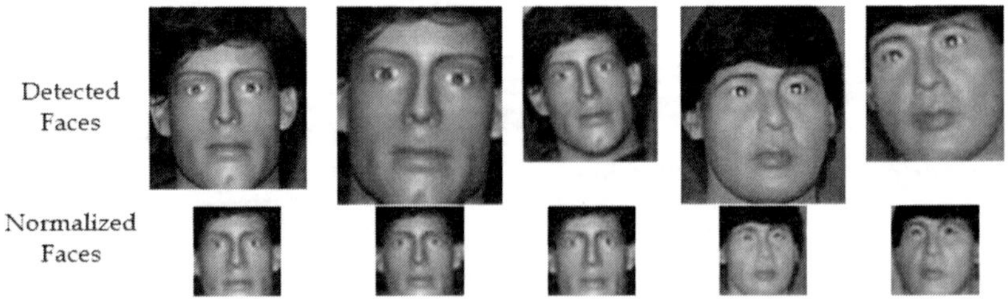

Figure 2. Geometric normalization (Before (top), After (bottom))

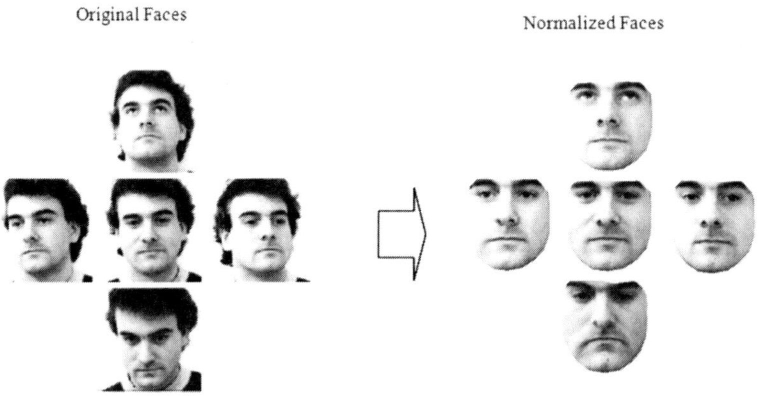

Figure 3. Complex geometric normalization

2.1.3. Illumination Normalization

If we can control the image capturing environment and impose strict requirements regarding lighting conditions (i.e. control access), recognition accuracy can be improved. In most scenarios where video face recognition is employed, the variations in lighting conditions when the faces are captured can range between dark and bright extremes. The profile face samples for each person to be recognized are captured in very different conditions with still images than those used in video imagery. A pre-processing algorithm should be used to minimize the effect of the lighting conditions when capturing the video images.

Depending on the resources available and capturing conditions, the illumination normalization algorithm can vary from simple algorithms such as: histogram equalization (HE), contrast limited adaptive histogram equalization (CLAHE) (Pizer et al., 1987; Corcoran et al., 2006a), logarithm transformed combined with suppressing DCT coefficients (LogDCT) or retinex (Land, 1986) based approaches, to more complex algorithms that can model the effects of lighting over facial regions (Lee et al., 2001; Smith & Hancock, 2005).

For embedded devices the simple normalization is a good compromise between execution speed and robustness to lighting variations.

Figure 4 displays the output of simple normalization techniques for two images affected by extreme side illumination.

An important issue to be considered when designing an illumination normalization algorithm is the balance between minimizing the effect of illumination and the inherent loss of information; information which is useful for classification. For instance, a face may appear dark because of dark lighting conditions or because the person has dark skin. Usually after normalization this information may not be recovered.

Figure 4. Simple illumination normalization examples

The validation sets for most of the algorithms that try to minimize the effect of lighting (i.e. Yale database (Georghiades et al.,2001)) consists of faces captured in very different lighting conditions where the normalization algorithms have better results compared with using the original faces without illumination normalization. In real life conditions, the faces can be compared to similar lighting conditions where applying the illumination normalization should not have a negative impact over the recognition results.

It is very important to have a validation set that has a variation distribution close to those most likely to be encountered in the scenarios that the recognition system is designed for.

2.1.4. Feature Extraction

Together with the useful information that can be used to differentiate between individuals, the face images described by the pixel values contain redundant information and information that can be ignored in the classification stage. By extracting only the useful information in this step we improve the accuracy of the recognition and also lower the storage requirement for each face.

Below are the main requirements for the feature selection algorithm:

- Good discriminative property. The features need to be able to differentiate between people. This translates into large variations between the value distributions for each person.

- Consistency. Features should not be modified between different images of the same person. This allows for recognition accuracy across large variations. Quantitatively this translates into small variation in the feature distribution for multiple faces of the same person.

- Small size. These features need to be stored and compared. Small size will allow fast comparison and low storage requirement.

- Fast computation. In order to achieve real-time recognition in video images, the faces need to be processed quickly.

The first two requirements will improve the accuracy of the recognition system and the last two requirements will ensure real-time recognition in embedded devices.

Classical approaches for still image recognition were also applied to the video image scenario with good results. These include: Principal Component Analysis (PCA) (Turk & Pentland, 1991), Linear Discriminate Analysis (LDA) (Belhumeur et al., 1996) and Discrete Cosine Transform (DCT) (Podilchuk & Zhang, 1996).

The DCT approach is of particular interest because of the speed of DCT transformation. Most of the capturing devices have DCT already implemented as part of JPEG compression module for storing captured images. More recent approaches like Local Binary Patterns (LBP) (Ojala et al., 2001) and Histogram of Oriented Gradients (HOG) (Lowe, 2004) have been used for face recognition.

2.1.5. Classification

In the case of still image recognition, the system makes a decision if the test face belongs to one of the people in the database and if so, which one (based on comparing the features computed in the previous step for test faces and a database of people).

In the case of video image recognition, the system compares the series of test faces with those in the sample database. Most commonly this is implemented as a series of still images derived comparisons and at each frame the confidence of our decision is modified based on the history of previous comparisons.

Simple classification algorithms like distance between feature vectors are preferred because of their simplicity and speed. More complex learning algorithms can be used if there are enough computation resources.

The classification algorithm is divided into two stages:

1 Training. Prototypes are constructed for each person in the database. The prototypes can be built from single or multiple face samples. Using multiple samples improves the quality of the prototype. The prototype can be represented by a series of feature vectors (such as distance-based classification) or can be represented by statistical models trained with multiple samples (such as learning-based classification algorithms).

2 Testing. Test samples are compared with each person prototype and similarity scores are computed. A decision is made using these similarity scores and the history of previous scores.

If, for a specific scenario, there is a fixed database that does not modify or update on the same platform where recognition is executed, a more complex algorithm for training can be used (i.e. training a learning algorithm) and performed offline. The result of this training algorithm is used during the recognition phase. When the database needs to be updated on-line at any time, the training algorithm needs to be less complex to be run on the embedded device. The result of the training algorithm, either the feature vectors or the person model, needs to be stored in the training database. This will influence the storage requirement of the recognition system.

An interesting algorithm that can be used for video face recognition is to model not only the appearance of the person at each frame but also the transition from frame to frame. A multi-dimensional Hidden Markov Model (HMM) (Nefian & Hayes, 1998) is used in order to model this type of transition. At the moment the complexity of HMMs makes it less favourable for embedded implementation.

2.2. Performance Testing

Comparing two face recognition systems is a difficult task because there are many parameters that can describe the performance of a particular recognition system. Usually one system or the other is superior using different sets of parameters.

Depending on the specific application where the recognition system is deployed, some specific performance metrics are more important than others. For example, a security system based on face recognition will have, as a main priority, a very low false acceptance rate, whereas a photo sorting application implemented on an embedded platform based on face recognition, will have its priorities of high recognition rates and low complexity.

The performance of a recognition system can be described by two types of parameters: accuracy parameters that describe how accurate the system is in recognizing faces and technical parameters that represent characteristics such as how fast the system will process a face, etc.

Some of the accuracy parameters that can be used to describe a video face recognition system include:

- Recognition rate. This is the main measurement to describe the accuracy of a recognition system. It represents how many faces are correctly recognized from the total number of faces. For video recognition this is a little more complex as it can be computed as the total number of frames where the faces are recognized.

- False positive rate. For specific applications this parameter can be more important than the recognition rate. This is usually computed as the number of mistakes made by the system. It can be further classified as a false acceptance rate in verification applications where an unknown individual is classified as one person from the database and as a false rejection rate where a person from the database is classified as unknown.

- Receiver Operating Characteristic (ROC) curve. In most cases there is a trade-off between the recognition rate and a false positive rate. For a high recognition rate, tune the recognition system to increase the recognition rate. This will inevitably increase the false positive rate as well and the other way around. The ROC curve represents the recognition rate for each possible false positive rate and only by displaying the ROC curves can a comparison of the two recognition algorithms be made.

- Minimum face size to be detected and recognized. When working with normal video resolution the faces can be very small, even at short distances from the capture equipment. Imposing a high minimum size for the face in order to be recognized can lead to a high rate of faces that are ignored or not recognized in the video images.

- Range of pose variations to recognize a face. Depending on the application to apply the recognition, a higher or lower range of pose variations is needed to recognize the faces.

- There are many technical parameters that can be used to describe a video face recognition system including:

- Processing time. This represents the time required to detect, process and classify all faces in a frame. This parameter depends on the platform where the recognition is implemented and will dictate if real-time functionality is available or not. For video frames, the time available for real-time recognition is the time between consecutive frames.

- Memory requirements. This represents the storage requirement for the system and includes the size of the feature vectors, person prototypes and other constants used in the algorithm.

- Number of faces recognized in each frame. The time required for detecting all faces in a frame is constant. This parameter will be influenced by the time required to process and recognize one face after it is detected.

The accuracy parameters depend on the database used for testing and the technical parameters depend on the specific platform where the recognition system is implemented. Without using the same database and same platform, two recognition systems cannot be compared only by the performance parameters.

3. PROPOSED RECOGNITION SYSTEM

The goal is to build a video face recognition system running in real-time on low computational power embedded platforms capable of recognizing multiple faces in video sequences. The main use case scenario intended for this system is tagging faces in consumer images as they are captured by digital cameras and mobile phones. The main requirements for the recognition system are high recognition rate, high robustness to variation types and low computational complexity for the algorithms used in the system. For this scenario, the input video stream can vary in size from small (qVGA) to high (full HD). Faces can also vary in size from tens of pixels in width to hundreds of pixels. Large variations in face pose, expression and illumination are also likely to be present.

Tessera's Face Tracking and Detecting technology is used in this experiment (Tessera, 2010). For geometric normalization, a computationally attractive approach is used, which involves: gray-scale transformation of the image, rotation of the face image to align the eyes on horizontal direction and resizing the face image to a small fixed size (i.e. 32x32 pixels). This size will allow for recognition of faces at a range of distances from the camera.

To minimize the effect of lighting variations, use a variant of the retinex (Land, 1986) illumination normalization algorithm. This is done by using a fixed variance matrix computed offline from a large database of images. This approach is very fast to apply and insures that the features computed in the next stage are more robust to large variations in illumination.

The features used for classification, in this chapter, are a variant of the Local Binary Pattern (LBPs) (Ojala et al., 2001) features which have been recently employed, with good results, for face recognition (Ahonen et al., 2006).

The classic approach of using LBP features in face recognition involves computing these features for each pixel in the face image, dividing the face image into small regions (separated or overlapped), and for each region computing the distribution of the LBP values. Often, only a small subset of all the features is used (uniform LBPs) in order to compute the region distribution. The classification involves comparing these distributions between corresponding regions from the test faces and the face samples used to build the prototypes in the training stage.

One approach is completely different. It is based on selecting from all possible features, those features that maximize the two properties defined in Section 2.1.4, namely: consistency and discriminancy. The training stage is split into two stages:1) Off-line training, using a very large database of faces, in order to determine and select the most consistent features. 2)On-line training, using the face samples in the database that need to be recognized.

The weights for each selected feature are computed in the off-line stage, each weight representing how discriminating the respective features are for the people in the database. Look for the best features that are globally consistent for a very large database (off-line) of people and weight them according to how well they can discriminate for a given database (on-line). Both training stages are presented in the next sections.

For classification a similarity measure between two faces is computed by looking for identical corresponding features in the off-line training stage. For each identical feature value add the similarity between the energy of the features multiplied by the weights computed in the on-line training stage.

3.1. Feature Extraction – Lbp Features

The Local Binary Patterns (LBP) (Ojala et al., 2001) features have been used in the system. These features are computed based on comparing the central pixel with its neighbours, concatenating the binary comparison results and computing the decimal number from the binary string.

Figure 5 shows how the feature is computed from an image.

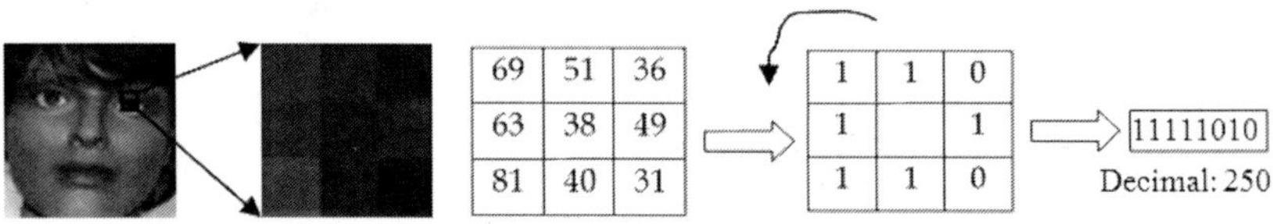

Figure 5. LBP feature.

The LBP features extract local information from the face region and due to the binary comparisons are robust to changes in illumination.

3.1.1. Multi-Scale Lbps and Extended Lbps

In order to capture more information from the face region, features are computed at different face resolutions beginning with the standard size used for geometrical normalization, down to smaller scales by downsampling the face image with different factors (i.e. 2, 4 etc). The LBP feature are extracted at each resolution using their 8 closest neighbours together with their extended variants using 8 more distant neighbours. Figure 6 illustrates an example of the normal LBP feature together with its first order extended LBP feature.

For each pixel in the normalized face image, compute multiple feature values. Do the same for the other scales.

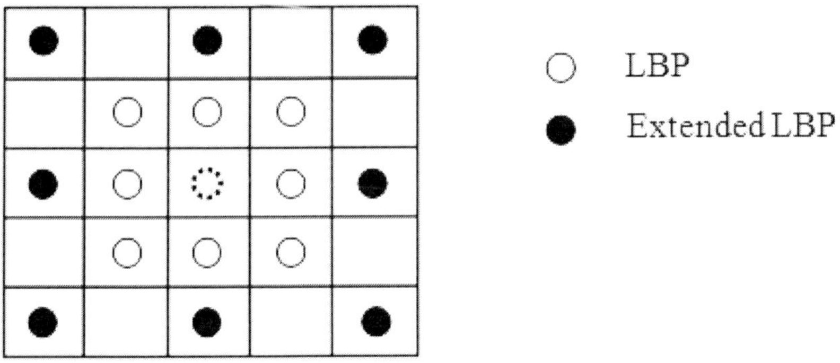

Figure 6. LBP and Extended LBP

3.1.2. Lbp Energy

The binary comparisons used for computing the features make them very robust to illumination variations. They also cause loss of information about the similarity of the local regions. For example, a very strong feature will have same value as a very faded feature. For this, calculate the normalized energy of the feature that will be used when comparing identical features for similarity between faces. The energy is computed using the formula:

$$e = \sum I_i^2 / max(I_i^2) \tag{1}$$

where I_i represents the value of the neighbour pixel i used when computing the feature. The energy is computed for both normal LBP and extended LBP.

Because small size face images are used, the features are not grouped in the face image by dividing the face into regions, but the corresponding features are compared for classification. The features are corresponding if they are computed at the same location, same scale and if they are normal or extended.

The feature vector after this analysis consists of the normal and extended LBPs computed at each location and at each scale together with their energies.

3.2. Off-Line Training – Consistency Analysis

Good features are those that do not change between images of the same person. This increases the accuracy of the recognition for different variations of the facial image.

This algorithm ranks a set of features given a large database of facial images. The order of the features is given considering their intra-class variation from low to high. The first features will be the most consistent between faces of same individuals. For recognition these features are more robust to variations.

Assume there is a collection of *m* people ($P\,1,\ P\,2,...P\,m$), each with multiple facial images. For each face compute all *N* possible features described in the previous section (normal and extended LBPs at all resolutions). Note the feature vector for person $P\,i$ image *j* as:

$$F_{ij} = (F_{ij1},\ F_{ij2}...\ F_{ijN}) \tag{2}$$

The features in the system are the LBP features. This algorithm can be extended to any other type of feature or combinations between features. Below is the general form of the algorithm.

For each feature, define a measure of intra-class consistency $S=(S\,1,S\,2,...,S\,N)$

The steps of the algorithm are:

- Reset all scores.

$$S = (0) \tag{3}$$

- Update scores. For every feature *k,* for every person *i,* for every *m* image of person *i:*

- Compare the feature $F\,imk$ with the same feature *k* of the remaining *j* images of person *i* ($F\,ijk$ with $j=1{:}N\,i$) where $N\,i$ is the number of images for person *i*.

- If $|F\,imk - F\,ijk| < thr$ then increment $S\,k$

$$S_k = S_k + 1 \tag{4}$$

At the end of this process order all features according to their score. Depending on the constraints, either keep a fixed number of features for classification in the latter stage or impose a threshold over the consistency measure.

The $|F\,imk - F\,ijk|$ term represents the distance between the feature values. In this case, search for identical features so $thr = 0$. For other types of features, a distance measure needs to be defined and a suitable *thr* needs to be chosen.

For best results, meaning best globally consistent features, the input database should be very large with all types of variation. Because it is executed off-line it does not affect the speed performance of the overall system.

3.3. On-Line Training – Discriminative Analysis

Together with consistency, the features also need to be able to discriminate between the people in the database. The same value for one feature across a database means both perfect consistency and no discriminative power. This algorithm assigns weights to each previously selected feature. The weights determine how well the feature can discriminate between the person in the database and which can be used in the classification stage.

Apply this algorithm to any type of feature using a suitable distance measure. The algorithm description is generic.

Assume *m* people in the database ($P1, P2, \ldots Pm$), each with at least one representative facial image. For each face, *N* representative features selected by the off-line training procedure and their corresponding discriminative scores. For example, for person Pi we have:

- feature vector $F_i=(F_{i1},F_{i2},\ldots,F_{iN})$
- score vector $S_i=(S_{i1},S_{i2},\ldots,S_{iN})$
- The steps of the algorithm are:
- Reset all scores.

$$S_i = (0),\ i = 0 \ldots m \qquad (5)$$

- Update scores. For every feature *k*, for every person *i*,
- Compare the feature $F\,ik$ with the same feature *k* of the remaining people ($F\,jk$ with $j=1:N$)
- If $|F_{ik} - F_{jk}| >$ thr then increment S_{ki}

$$S_{ki} = S_{ki}+1 \text{ or or } S_{ki} = S_{ki} + |F_{ik} - F_{jk}| \qquad (6)$$

In order to make these scores independent of the number of people and faces in the database, normalize them using the maximum sum of scores for a person using the next equation where *N* represents the number of features and *m* the number of people in the database:

$$S_{ki} = S_{ki}/max(sum(S_{ln})_{n=1:N})_{l=1:M} \quad (7)$$

The same observations from the previous section for the terms: $|F_{ik} - F_{jk}|$ and *thr* are valid. In the case of searching for identical LBPs, the parameter *thr = 0.*

3.4. Classification

Having computed the features described in Section 3.1 and the discriminative scores of all features selected in Section 3.2, using the algorithm described in the previous section, compute a similarity (S_{ij}) between two faces (trained face f_i and test face f_j) by counting how many identical corresponding features there are between the two faces using this formula:

$$S_{ij} = \sum w_k * | e_{ik} - e_{jk} | * g_{ij}, k=1:N \quad (8)$$

where:

- g_{ijk} is equal to 1 if feature *k* is identical between faces *i* and *j,* and 0 otherwise.
- e_{ik}, e_{jk} represent the energy of the feature *k* from image *i* and *j* respectively computed using eq. (1).
- w_k represents the discriminative score for features *k* of trained face *i.*

Comparing the test face with all face samples from the trained database, return the most similar person with the test face. By imposing a decision threshold, control the recognition rate versus the false positive rate, depending on the application mentioned in Section 2.2. Once the similarity measure between the test face and the most similar person from the database is higher than the decision threshold, decide that the face is recognized and continue or not the recognition process over the next frames.

4. RESULTS

In order to assess the performance of the recognition system a large database of videos with systematic variations was used, including: pose, illumination, face size/distance to subject, and facial expressions.

In the training stage, a single image was used to train each person. The training face is frontal, good size, normal illumination and good quality. Tests were run for different numbers of people in the database from low (3) to high (100).

For each test, the recognition rate (RR) was measured, as the number of correct classifications, false positive rate (FP2) as wrong classifications and undecided rate (MD) as number of test faces which were not classified.

Figure 7 shows the recognition and error rates as a function of head yaw angle. As specified above, training was conducted at head yaw angles of zero degrees and testing was done with 0°, 10°, 20° and 30° yaw angles.

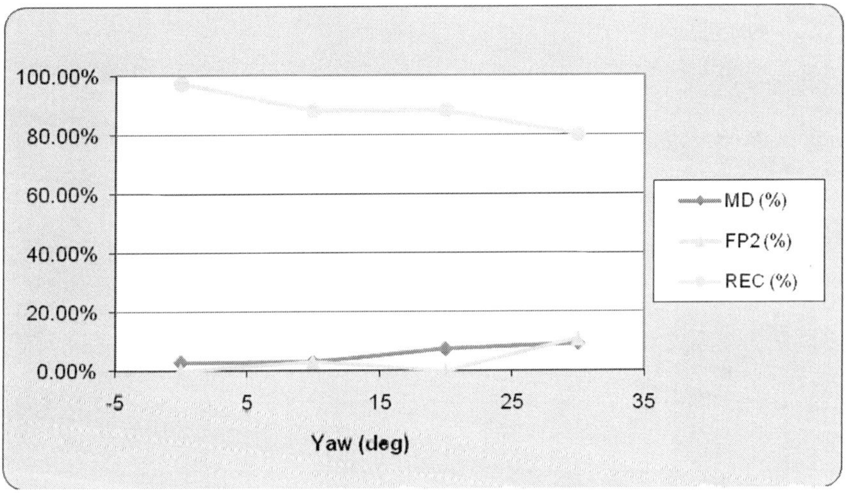

Figure 7. Recognition performance for different yaw angles

Figure 8 shows the recognition and error rates as a function of head pitch angle. Training was conducted at head pitch angles of zero degrees.

Figure 9 shows the recognition and error rates as a function of different facial expressions. The training faces had no facial expression.

Figure 10 shows the recognition and error rates as a function of different illumination conditions. The approximate EV values for the given conditions are: LowLight (2.4EV) and StrongLight (9EV), which can be considered extreme lighting conditions. Training was conducted using normal indoors ambient lighting.

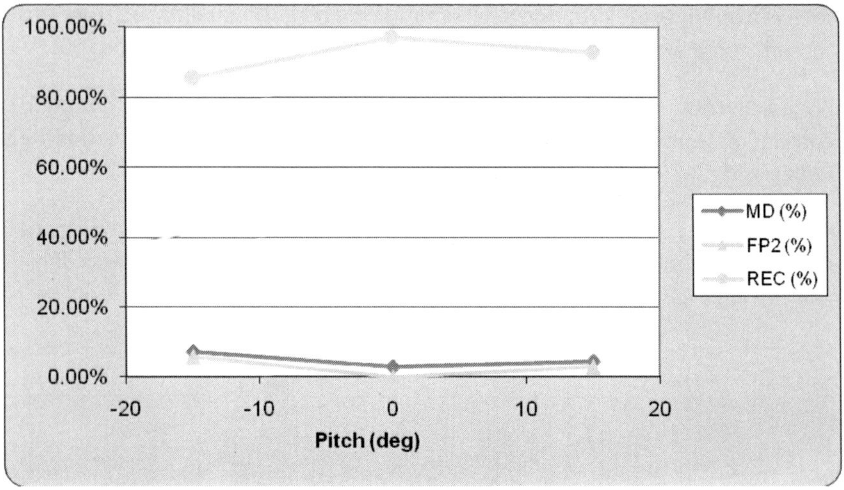

Figure 8. Recognition performance for different pitch angles

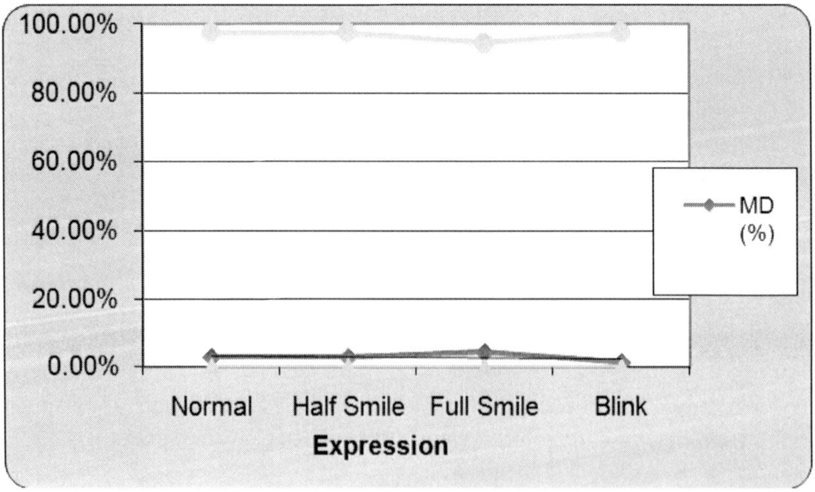

Figure 9. Recognition performance for different face expressions

The main technical parameters for the system were implemented on an ARM9 platform (266 MHz CPU), the processing time for a face depending on the size of the input frame varied between 8 and 15 milliseconds for qVGA and VGA input frame size which is well within real-time requirements. The size of the features vector for each analyzed face is about 2Kb which is very small.

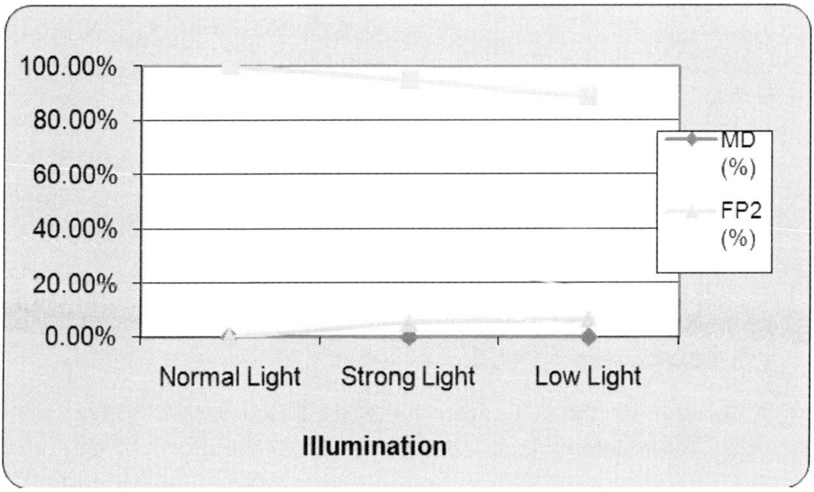

Figure 10. Recognition performance for illumination conditions

5. CONCLUSION

This chapter presented the challenges of implementing a real-time video face recognition system on an embedded platform. The first section presented the main issues that need to be addressed when designing such a system and possible solutions. The second part described a working solution based on using LBP features which are fast to compute, robust to variations and able to extract useful information from the face region. In order to obtain a robust recognition system, only the features which have the same value across multiple variations of the same person were extracted. In order to increase the accuracy of the system, weights were associated to the selected features based on their discriminative power between the people from the database.

Results for this system were tested and implemented on an embedded platform, which shows good accuracy across large variations of the input data and technical parameters which satisfy the condition for real-time processing.

REFERENCES

1. T. Ahonen, A. Hadid, M. Pietikainen, 2006 Face description with local binary patterns: Application to face recognition. IEEE Transactions on Pattern Analysis and Machine Intelligence PAMI, 28, 2037 2041 , December 2006.

2. P. N. Belhumeur, J. P. Hespanha, D. J. Kriegman, 1996 Eigenfaces vs. Fisherfaces: Recognition using Class Specific Linear Projection, Proc. of the 4th European Conference on Computer Vision, ECCV'96, 15-18 April 1996, Cambridge, UK, 45 58

3. P. Corcoran, G. Costache, 2005 Automated sorting of consumer image collections using face and peri heral region image classifiers , Consumer Electronics, IEEE Transactions on 51 3 Aug. 2005, 747 754 .

4. P. Corcoran, C. Iancu, G. Costache, 2006a Improved hmm based face recognition system. International Conference on Optimization of Electrical and Electronic Equipment, Brasov, Romania, May 2006.

5. P. Corcoran, M. Ionita, G. Costache, 2006b Pose-invariant face recognition using AAMs, International Conference on Optimization of Electrical and Electronic Equipment, Brasov, Romania, May 2006.

6. G. Costache, R. Mulryan, E. Steinberg, P. Corcoran, 2006 In-camera person-indexing of digital images, Consumer Electronics ICCE '06 Digest of Technical Papers. International Conference on, 7 11 Jan. 2006.

7. S. Georghiades, P. N. Belhumeur, D. J. Kriegman, 2001 From few to many: illumination cone models for face recognition under variable lighting and pose, IEEE Trans. Pattern Anal. Mach. Intelligence, 23 6 643 660 .

8. Google Picassa (n.d.), http://picasa.google.com/

9. Land, 1986 An alternative technique for the computation of the designator in the retinex theory of color vision. Proc. Nat. Acad. Sci., 83 3078-3080

10. K. Lee, J. Ho, D. Kriegman, 2001 Nine points of light: Acquiring subspaces for face recognition under variable lighting. In Proc. of CVPR, 519 526 .

11. D. G. Lowe, 2004 Distinctive image features from scale-invariant keypoints. 226 International Journal of Computer Vision 60 (2), 91 EOF 110 EOF .

12. V. Nefian, I. I. I. M. H. Hayes, 1998 Hidden Markov Models for Face Recognition, Proc. of the IEEE International Conference on Acoustics,

Speech, and Signal Processing, ICASSP'98, 5 12-15 May 1998, Seattle, Washington, USA, 2721 2724

13. T. Ojala, M. Pietikäinen, T. Mäenpää, 2001 A generalized Local Binary Pattern operator for multiresolution gray scale and rotation invariant texture classification, Advances in Pattern Recognition, ICAPR 2001 Proceedings, Springer, 397 406

14. P. J. Phillips, P. Grother, R. Micheals, D. M. Blackburn, E. Tabassi, M. Bone, 2003 Face Recognition Vendor Test 2002, Proceedings of the IEEE International Workshop on Analysis and Modeling of Faces and Gestures, 44 October 17, 2003

15. S. Pizer, E. Amburn, J. Austin, R. Cromartie, A. Geselowitz, T. Greer, B. Romeny, J. Zimmerman, K. Zuiderveld, 1987 Adaptive histogram equalization and its variations. Computer Vision, Graphics, and Image Processing, 39 355 368 .

16. C. Podilchuk, X. Zhang, 1996 Face recognition using DCT-based feature vectors, Proc. IEEE International Conference on Acoustics, Speech and Signal Processing (ICASSP'96), 4 2144 2147 , May 1996.

17. W. A. P. Smith, E. R. Hancock, 2005 Single image estimation of facial albedo maps. BVAI, 517 526 .

18. Tessera OptiML FaceTools (2010). http://tessera.com/technologies/ imagingandoptics/ \Documents/OptiML_faceTools.pdf

19. M. A. Turk, A. P. Pentland, 1991 Face Recognition using Eigenfaces, Proc. of IEEE Conference on Computer Vision and Pattern Recognition, 586 591 , June 1991.

20. OptiML is a trademark of Tessera Inc. In the United States and other countries.

Chapter 8

Comparison of Spectral-Only and Spectral/Spatial Face Recognition for Personal Identity Verification

Zhihong Pan,1 Glenn Healey,2 and Bruce Tromberg3

[1] Galileo Group Inc., 100 Rialto Place Suite 737, Melbourne, FL 32901, USA

[2]Department of Electrical Engineering and Computer Science, University of California, Irvine, CA 92697, USA

[3]Beckman Laser Institute, 1002 East Health Sciences Road, Irvine, CA 92612, USA

ABSTRACT

Face recognition based on spatial features has been widely used for personal identity verification for security-related applications. Recently, near-infrared spectral reflectance properties of local facial regions have been shown to be sufficient discriminants for accurate face recognition. In this paper, we compare the performance of the spectral method with face recognition using the eigenface method on single-band images extracted from the same hyperspectral image set. We also consider methods that use multiple original and PCA-transformed bands. Lastly, an innovative spectral eigenface method which uses both spatial and spectral features is proposed to improve the quality of the spectral features and to reduce the expense of the computation. The algorithms are compared using a consistent framework.

1. INTRODUCTION

Automatic personal identity authentication is an important problem in security and surveillance applications, where physical or logical access to locations, documents, and services must be restricted to authorized persons. Passwords or personal identification numbers (PINs) are often assigned to individuals for authentication. However, the password or PIN is vulnerable to unauthorized exploitation and can be forgotten. Biometrics, on the other hand, use personal intrinsic characteristics which are harder to compromise and more convenient to use. Consequently, the use of biometrics has been gaining acceptance for various applications. Many different sensing modalities have been developed to verify personal identities. Fingerprints are a widely used biometric. Iris recognition is an emerging technique for personal identification which is an active area of research. There are also studies to use voice and gait as primary or auxiliary means to verify personal identities.

Face recognition has been studied for many years for human identification and personal identity authentication and is increasingly used for its convenience and noncontact measurements. Most modern face recognition systems are based on the geometric characteristics of human faces in an image [1–4]. Accurate verification and identification performance has been demonstrated for these algorithms based on mug shot type photographic databases of thousands of human subjects under controlled environments [5, 6]. Various 3D face models [7, 8] and illumination models [9, 10] have been studied for pose and illumination-invariant face recognition. In addition to methods based on gray-scale and color face images over the visible spectrum, thermal infrared face images [11, 12] and hyperspectral face images [13] have also been used for face recognition experiments. An evaluation of different face recognition algorithms using a common dataset has been of general interest. This approach provides a solid basis to draw conclusions on the performance of different methods. The Face Recognition Technology (FERET) program [5] and the Face Recognition Vendor Test (FRVT) [6] are two programs which provided independent government evaluations for various face recognition algorithms and commercially available face recognition systems.

Most biometric methods, including face recognition methods, are subject to possible false acceptance or rejection. Although biometric information is difficult to duplicate, these methods are not immune to forgery, or so-called spoofing. This is a concern for automatic personal identity authentication since intruders can use artificial materials or objects to gain unauthorized access. There are reports showing that fingerprint sensor devices have been deceived by Gummi fingers in Japan [14] and fake latex fingerprints in Germany [15]. Face and iris recognition systems can also be compromised since they use external observables [16]. To counter this vulnerability, many biometric systems employ a liveness detection function to foil attempts at biometric forgery [17, 18]. To improve system accuracy, there is strong interest in research to combine multiple biometric characteristics for multimodal personal identity authentication [19, 20]. Since hyperspectral sensors capture spectral and spatial information they provide the potential for improved personal identity verification.

Methods that have been developed consider the use of representations for visible wavelength color images for face recognition [21, 22] as well as the combination of color and 3D information [23]. In this work, we examine the use of combined spectral/spatial information for face recognition over the near-infrared (NIR) spectral range. We show that the use of spatial information can be used to improve on the performance of spectral-only methods [13]. We also use a large NIR hyperspectral dataset to show that the choice of spectral band over the NIR does not have a significant effect on the performance of single-band eigenface methods. On the other hand, we show that band selection does have a significant effect on the performance of multiband methods. In this paper we develop a new representation called the spectral-face which preserves both high-spectral and high-spatial resolution. We show that the spectral eigenface representation outperforms single-band eigenface methods and has performance that is comparable to multiband eigenface methods but at a lower computational cost.

2. FACE RECOGNITION IN SINGLE-BAND IMAGES

A hyperspectral image provides spectral information, normally in radiance or reflectance, at each pixel. Thus, there is a vector of values for each pixel corresponding to different wavelengths within the sensor spectral range. The reflectance spectrum of a material remains constant in different images while different materials exhibit distinctive reflectance properties due to different absorbing and scattering characteristics as a function of wavelength. In the spatial domain, there are several gray-scale images that represent the hyperspectral imager responses of all pixels for a single spectral band. In a previous study [24], seven hyperspectral face images were collected for each of 200 human subjects. These images have a spatial resolution of 468×494 and 31 bands with band centers separated by 0.01 μm over the near-Infrared (0.7 μm–1.0 μm). Figure 1 shows calibrated hyperspectral face images of two subjects at seven selected bands which are separated by 0.06 μm over 0.7 μm–1.0 μm. We see that the ratios of pixel values on skin or hair between different bands are dissimilar for the two subjects. That is, they have unique hyperspectral signatures for each tissue type. Based on these spectral signatures, a Mahalanobis distance-based method was applied for face recognition tests and accurate face recognition rates were achieved. However, the performance was not compared with classic face recognition methods using the same dataset.

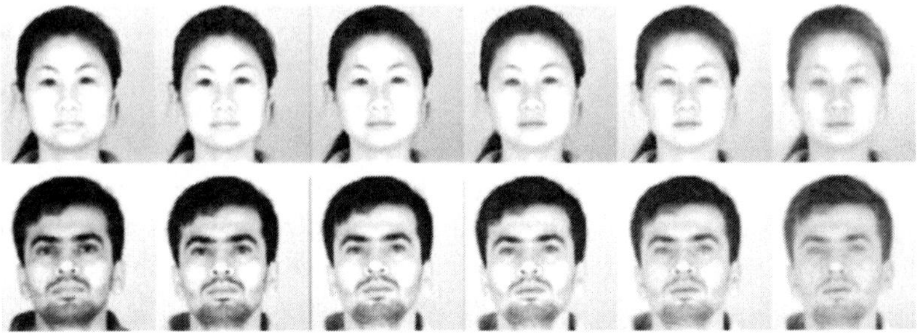

Figure 1. Selected single-band images of two subjects.

The CSU Face Identification Evaluation System [25] provides a standard set of well-known algorithms and established experimental protocols for evaluating face recognition algorithms. We selected the Principal Components Analysis (PCA) Eigenfaces [26] algorithm and used cumulative match scores as in the FERET study [5] for performance comparisons. To prepare for the face recognition tests, a gray-scale image was extracted for each of the 31 bands from a hyperspectral image. The coordinates of both eyes were manually positioned before processing by the CSU evaluation programs. In the CSU evaluation system all images were transformed and normalized so that they have a fixed spatial resolution of 130×150 pixels and the eye coordinates are the same. Masks were used to void nonfacial features. Histogram equalization was also performed on all images before the face recognition tests were conducted. For each of the 200 human subjects, there are three front-view images with the first two (fg and fa) having neutral expression and the other (fb) having a smile. All 600 images were used to generate the eigenfaces. Figure 2 shows one single-band image before and after the normalization, and the first 10 eigenfaces for the dataset. The number of eigenfaces used for face recognition was determined by selecting the set of most significant eigenfaces which account for 90% of the total energy.

Given the w th band of hyperspectral images U and V , the Mahalanobis Cosine distance [27] is used to measure the similarity of the two images. Let $u_{w,i}$ be the projection of the w th band of U onto the i th eigenface and let $\sigma_{w,i}$ be the standard deviation of the projections from all of the w th band images onto the i th eigenface. The Mahalanobis projection of U_w is $M_w = (m_{w,1}, m_{w,2}, \ldots, m_{w,I})$ where $m_{w,i} = u_{w,i} / \sigma_{w,i}$. Let N_w be the similarly computed Mahalanobis projection of V_w. The Mahalanobis Cosine distance between U and V for the w th band is defined by

$$D_{U,V}(w) = -\frac{M_w \cdot N_w}{|M_w||N_w|}, \tag{1}$$

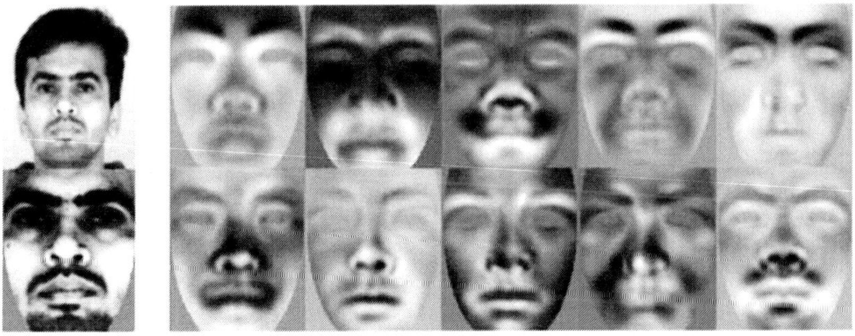

Figure 2. Example of eigenfaces in one single-band.

which is the negative of the cosine between the two vectors. For the 200 subjects, the fg images were grouped in the gallery set and the fa and fb images were used as *probes* [5]. The experiments follow the *closed universe* model where the subject in every image in the probe set is included in the gallery. For each probe image, the Mahalanobis Cosine distance between the probe and all gallery images is computed. If the correct match is included in the group of gallery images with the N smallest distances, we say that the probe is correctly matched in the top N. The cumulative match score for a given N is defined as the fraction of correct matches in the top N from all probes. The cumulative match score for $N = 1$ is called the recognition rate. Figure 3 plots the cumulative match scores for $N = 1, 5$, and 10 respectively. Band 1 refers to the image acquired at 700 nm and band 31 refers to the image acquired at 1000 nm. We see that all bands provide high recognition rates, with more than 96% of the probes correctly identified for $N = 1$ and over 99% for $N = 10$. It is important to consider the statistical significance of the results. For this purpose, we model the fraction of the probes that are correctly identified by a binomial distribution with a mean given by the measured identification rate p. The variance σ^2 of p is given by $400p(1-p)$ where 400 is the number of probes [28]. For an identification rate of 0.97 we have $\sigma = 3.4$ which corresponds to a standard deviation in the identification rate of 0.009 and for an identification rate of 0.99 we have $\sigma = 1.99$ which corresponds to a standard deviation in the identification rate of 0.005. Thus, for each of the three curves plotted in Figure 3 the variation in performance across bands is not statistically significant. Figure 4 compares the cumulative match scores using the spectral signature method [13] and the single-band eigenface method using the most effective band. We see that the spectral signature method performs well but somewhat worse than the best single-band method for matches with N less than 8. For $N = 1$, a recognition rate of 0.92 corresponds to a standard deviation in the recognition rate of 0.014 which indicates that the difference between the two methods in Figure 4 is statistically significant. The advantage of the spectral methods is pose invariance which was discussed in a previous work [13] but which is not considered in this paper.

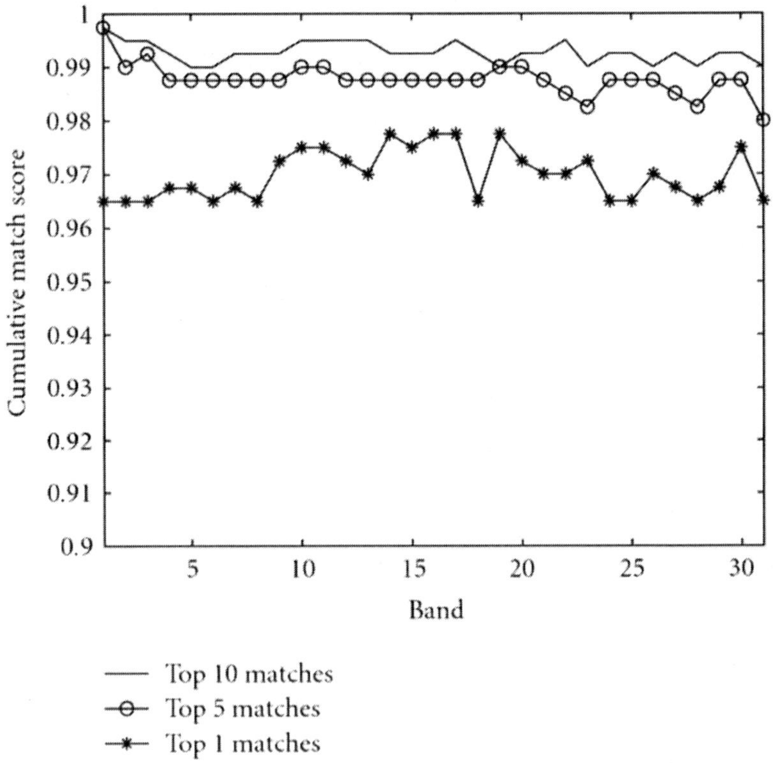

Figure 3. Cumulative match scores of single-band images at different wavelengths.

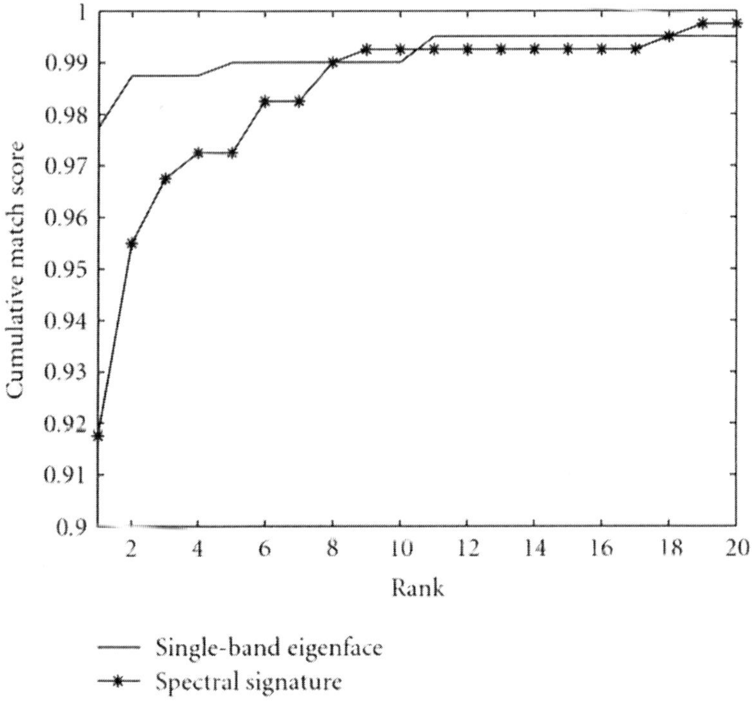

Figure 4. Cumulative match scores of spectral signature method and the best single-band eigenface method.

3. FACE RECOGNITION IN MULTIBAND IMAGES

We have shown that both spatial and spectral features in hyperspectral face images provide useful discriminants for recognition. Thus, we can consider the extent of performance improvements when both features are utilized. We define a distance between images U and V using

$$D_{U,V} = \sqrt{\sum_{w=1}^{W} (1 + D_{U,V}(w))^2}, \tag{2}$$

where the index w takes values over a group of W -selected bands that are not necessarily contiguous. Note that the additive 1 is to ensure a nonnegative value before the square.

Redundancy in a hyperspectral image can be reduced by a Principal Component Transformation (PCT) [29]. For a hyperspectral image $U = (U_1, U_2, \ldots, U_W)$, the PCT generates $U' = (U'_1, U'_2, \ldots, U'_W)$, where $U'_i = \Sigma_j \varepsilon_{ij} U_j$. The principal components $U'_1, U'_2, \ldots, U'_W$ are orthogonal to each other and sorted in order of decreasing modeled variance. Figure5 shows a single-band image at 700 nm and the first five principal components that are extracted from the corresponding hyperspectral image. We see that the first principal component image resembles the single-band image while the second and third component images highlight features of the lips and eyes. We also see that there are few visible features remaining in the fourth and fifth principal components.

Figure 5. Five principal band images of one subject after PCT.

Figure 6 plots the recognition rates for different multiband eigenface methods. First we selected the bands in order of increasing center wavelength and performed eigenface recognition tests for the first one band, two bands and up to 31 bands, respectively. We also sorted all 31 bands in descending order of recognition rate and performed the same procedure for the face recognition tests. From Figure 6 we see that both methods reach a maximum recognition rate of 98% when using multiple bands. However, when the number of bands is less than 16, the multiband method performs better if the bands are sorted in advance from the highest recognition rate to the lowest. We also used the leading principal components for multiband recognition. We see in Figure 6 that over 99% of the probes were correctly recognized when using the first three principal bands. Increasing the number of principal bands beyond 3 causes performance degradation. The original-order algorithm in Figure 6 achieves a recognition rate of approximately 0.965 for less than ten bands which corresponds to a standard deviation in recognition rate of 0.009. Thus, the performance difference between this method and the PCT-based method is significant between 3 and 9 bands. Note that the PCT was performed on each hyperspectral image individually with different sets of ε_{ij}. The PCT can also be implemented using the same coefficients for faster computation.

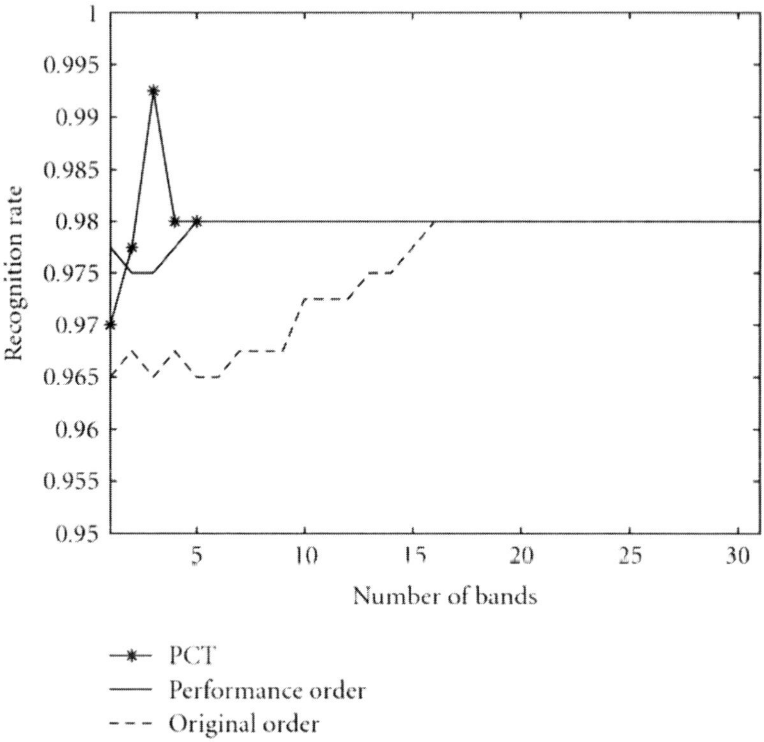

Figure 6. Recognition rate of multiband eigenface methods.

Figure 7 also compares the recognition performance of the three multiband methods discussed in the previous paragraph where each algorithm uses only the first three bands. It is interesting that sorting the bands according to performance improves the recognition rate for $N = 1$ but worsens the performance somewhat for larger values of N. In either case, the multiband method based on the PCT has the best performance for $N < 7$ and is equivalent to the original-order method for larger values of N.

4. FACE RECOGNITION USING SPECTRAL EIGENFACES

We showed in Section 3 that multiband eigenface methods can improve face recognition rates. In these algorithms, the multiple bands are processed independently. A more general approach is to consider the full spectral/spatial structure of the data. One way to do this is to apply the eigenface method to large composite images that are generated by concatenating the 31 single-band images. This approach, however, will significantly increase the computational cost of the process. An alternative is to subsample each band of the hyperspectral image before concatenation into the large composite image. For example, Figure 8 shows a 31-band image after subsampling so that the total number of pixels is equivalent to the number of pixels in a 130×150 pixel single-band image. We see that significant spatial detail is lost due to the subsampling.

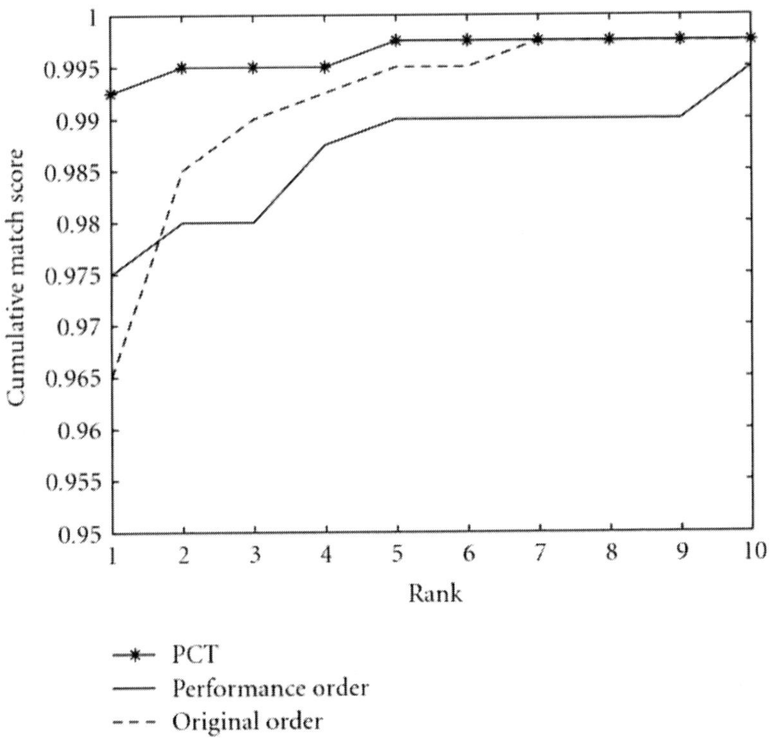

Figure 7. Cumulative match scores of multiband eigenface methods.

A new representation, called spectral-face, is proposed to preserve both spectral and spatial properties. The spectral-face has the same spatial resolution as a single-band image so the spatial features are largely preserved. In the spectral domain, the pixel values in the spectral-face are extracted sequentially from band 1 to band 31 then from band 1 again. For example, the value of pixel i in spectral-face equals the value of pixel i in band w where w is the remainder of i divided by 31. Figure 9 shows an original single-band image together with the normalized spectral-face image in the left column. Spectral-face has improved spatial detail as compared with Figure 8. The pattern on the face in Figure 9 demonstrates the variation in the spectral domain. With the spectral-face images, the same eigenface technique is applied for face recognition. The first 10 spectral eigenfaces are shown on the right side of Figure 9. It is interesting to observe that the eighth spectral eigenface highlights the teeth feature in smiling faces.

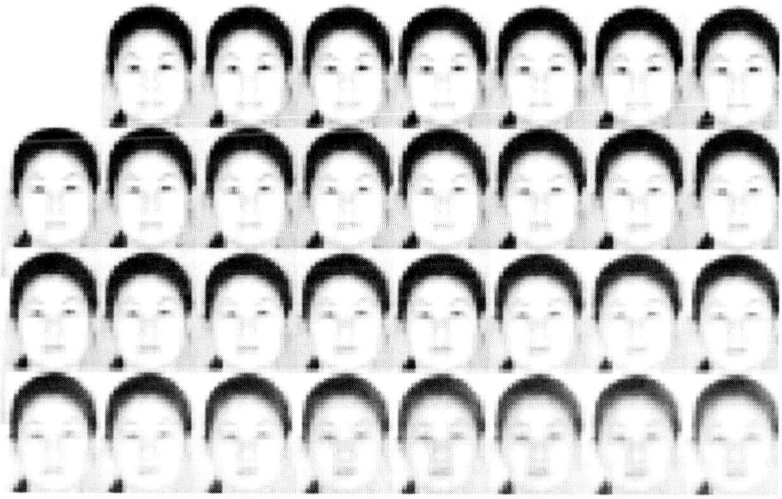

Figure 8. A sample image composed from 31 bands with low-spatial resolution.

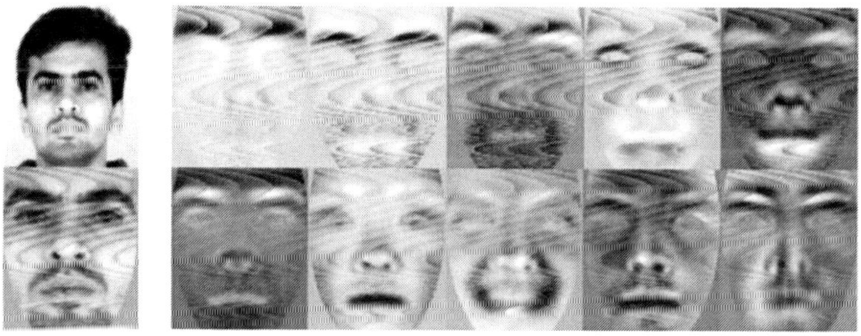

Figure 9. One sample spectral-face and the first 10 spectral eigenfaces.

The spectral eigenface method was applied to the same dataset as the single-band and multiband methods. The cumulative match scores for $N = 1$ to 20 are shown in Figure 10. The best of the single-band methods, which corresponds to band 19 (880 nm), is included for performance comparison with the spectral eigenface method. We see that the spectral eigenface method has better performance for all ranks. The best of the multiband methods, which combines the first three principal bands, is also considered. The multiband method performs better than the spectral eigenface method for small values of the rank, but performs worse for larger values of the rank. For this case, an identification rate of 0.99 corresponds to a standard deviation in identification rate of 0.005. Thus, the two multiple-band methods have a statistically significant advantage over the single-band eigenface method for ranks between 3 and 10. Note that the multiple principal band method requires more computation than the spectral eigenface method.

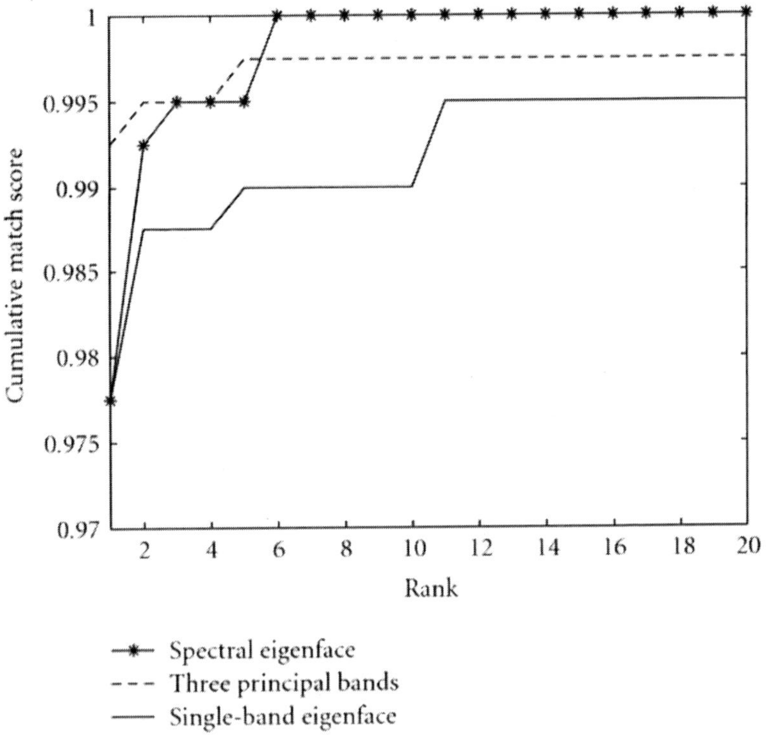

Figure 10. Comparison of spectral eigenface method with single-band and multiband methods.

5. CONCLUSION

Multimodal personal identity authentication systems have gained popularity. Hyperspectral imaging systems capture both spectral and spatial information. The previous work [24] has shown that spectral signatures are powerful discriminants for face recognition in hyperspectral images. In this work, various methods that utilize spectral and/or spatial features were evaluated using a hyperspectral face image dataset. The single-band eigenface method uses spatial features exclusively and performed better than the pure spectral method. However, the computational requirements increase significantly for eigenface generation and projection. The recognition rate was further improved by using multiband eigenface methods which require more computation. The best performance was achieved with the highest computational complexity by using principal component bands. The spectral eigenface method transforms a multiband hyperspectral image to a spectral-face image which samples from all of the bands while preserving spatial resolution. We showed that this method performs as well as the PCT-based multiband method but with a much lower computational requirement.

ACKNOWLEDGMENTS

This work was conducted when the author was with the Computer Vision Laboratory at the University of California, Irvine, USA. This work has been supported by the DARPA Human Identification at a Distance Program through AFOSR Grant F49620-01-1-0058. This work has also been supported by the Laser Microbeam and Medical Program (LAMMP) and NIH Grant RR01192. The data was acquired at the Beckman Laser Institute on the UC Irvine campus. The authors would like to thank J. Stuart Nelson and Montana Compton for their valuable assistance in the process of IRB approval and human subject recruitment.

REFERENCES

1. Chellappa R, Wilson CL, Sirohey S: **Human and machine recognition of faces: a survey.** *Proceedings of the IEEE* 1995,**83**(5):705-740.

2. Etemad K, Chellappa R: **Discriminant analysis for recognition of human face images.** *Journal of the Optical Society of America A* 1997,**14**(8):1724-1733.

3. Moghaddam B, Pentland A: **Probabilistic visual learning for object representation.** *IEEE Transactions on Pattern Analysis and Machine Intelligence* 1997,**19**(7):696-710.

4. Wiskott L, Fellous J-M, Krüger N, von der Malsburg C: **Face recognition by elastic bunch graph matching.***IEEE Transactions on Pattern Analysis and Machine Intelligence* 1997,**19**(7):775-779.

5. Phillips PJ, Moon H, Rizvi SA, Rauss PJ: **The FERET evaluation methodology for face-recognition algorithms.** *IEEE Transactions on Pattern Analysis and Machine Intelligence* 2000,**22**(10):1090-1104.

6. Phillips PJ, Grother P, Micheals R, Blackburn DM, Tabassi E, Bone M: *Face recognition vendor test 2002: overview and summary.* Defense Advanced Re-search Projects Agency, Arlington, Va, USA; March 2003.

7. Blanz V, Vetter T: **Face recognition based on fitting a 3D morphable model.** *IEEE Transactions on Pattern Analysis and Machine Intelligence* 2003,**25**(9):1063-1074.

8. Chang KI, Bowyer KW, Flynn PJ: **An evaluation of multimodal 2D+3D face biometrics.** *IEEE Transactions on Pattern Analysis and Machine Intelligence* 2005,**27**(4):619-624.

9. Adini Y, Moses Y, Ullman S: **Face recognition: the problem of compensating for changes in illumination direction.** *IEEE Transactions on Pattern Analysis and Machine Intelligence* 1997,**19**(7):721-732.

10. Lee K-C, Ho J, Kriegman DJ: **Acquiring linear subspaces for face recognition under variable lighting.***IEEE Transactions on Pattern Analysis and Machine Intelligence* 2005,**27**(5):684-698.

11. Socolinsky DA, Selinger A, Neuheisel JD: **Face recognition with visible and thermal infrared imagery.***Computer Vision and Image Understanding* 2003,**91**(1-2):72-114. 10.1016/S1077-3142(03)00075-4

12. Wilder J, Phillips PJ, Jiang C, Wiener S: **Comparison of visible and infra-red imagery for face recognition.***Proceedings of the 2nd International Conference on Automatic Face and Gesture Recognition (AFGR '96), October 1996, Killington, Vt, USA* 182-187.

13. Pan Z, Healey G, Prasad M, Tromberg B: **Face recognition in hyperspectral images.** *IEEE Transactions on Pattern Analysis and Machine Intelligence* 2003,**25**(12):1552-1560.

14. Leyden J: **Gummi bears defeat fingerprint sensors.** *The Register* 2002.

15. Harrison A: **Hackers claim new fingerprint biometric attack.** *Security Focus* 2003.

16. Lewis M, Statham P: **CESG biometric security capabilities programme: method, results and research challenges.** *Biometric Consortium Conference, September 2004, Crystal City, Va, USA*

17. Bigun J, Fronthaler H, Kollreider K: **Assuring liveness in biometric identity authentication by real-time face tracking.** *Proceedings of IEEE International Conference on Computational Intelligence for Homeland Security and Personal Safety (CIHSPS '04), July 2004, Venice, Italy* 104-111.

18. Tan T, Ma L: **Iris recognition: recent progress and remaining challenges.** *Biometric Technology for Human Identification, April 2004, Orlando, Fla, USA, Proceedings of SPIE* **5404:** 183-194.

19. Kittler J, Matas J, Jonsson K, Ramos Sánchez MU: **Combining evidence in personal identity verification systems.** *Pattern Recognition Letters* 1997,**18**(9):845-852.

20. Kittler J, Messer K: **Fusion of multiple experts in multimodal biometric personal identity verification systems.** *Proceedings of the 12th IEEE Workshop on Neural Networks for Signal Processing, December 2002, Kauai, Hawaii, USA* 3-12.

21. Yang J, Zhang D, Xu Y, Yang J-Y: **Recognize color face images using complex eigenfaces.** *Proceedings of International Conference on Advances in Biometrics (ICB '06), January 2006, Hong Kong, Lecture Notes in Computer Science* **3832:** 64-68.

22. Yoo S, Park R-H, Sim D-G: **Investigation of color spaces for face recognition.** *Proceedings of IAPR Conference on Machine Vision Applications (MVA '07), May 2007, Tokyo, Japan* 106-109.

23. Tsalakanidou F, Tzovaras D, Strintzis MG: **Use of depth and colour eigenfaces for face recognition.** *Pattern Recognition Letters* 2003,**24**(9-10):1427-1435.

24. Pan Z, Healey G, Prasad M, Tromberg B: **Face recognition in hyperspectral images.** In *Proceedings of IEEE Computer Society Conference on Computer Vision and Pattern Recognition (CVPR '03), June 2003, Madison, Wis, USA. Volume 1*. Institute of Electrical and Electronics Engineers; 334-339.

25. Bolme D, Beveridge JR, Teixeira M, Draper BA: **The CSU face identification evaluation system: its purpose, features and structure.** *Proceedings of the 3rd International Conference Computer Vision Systems (ICVS '03), April 2003, Graz, Austria, Lecture Notes in Computer Science* **2626:** 304-313.

26. Turk MA, Pentland AP: **Face recogntion using eigenfaces.** *Proceedings of IEEE Computer Society Conference on Computer Vision and Pattern Recognition (CVPR '91), June 1991, Maui, Hawaii, USA* 586-591.

27. Beveridge JR, Bolme DS, Teixeira M, Draper B: *The CSU face identification evaluation system user's guide: version 5.0.* Computer Science Department, Colorado State University, Fort Collins, Colo, USA; May 2003.

28. Papoulis A: *Probability and Statistics*. Prentice-Hall, Englewood Cliffs, NJ, USA; 1990.

29. Ready PJ, Wintz PA: **Information extraction, SNR improvement, and data compression in multispectral imagery.** *IEEE Transactions on Communications* 1973,**21**(10):1123-1131.

Chapter 9

A Cognitively-Motivated Framework for Partial Face Recognition in Unconstrained Scenarios

João C. Monteiro * and Jaime S. Cardoso

INESC TEC and Faculdade de Engenharia, Universidade do Porto,
Campus da FEUP, Rua Dr. Roberto Frias, n 378, 4200-465 Porto, Portugal

ABSTRACT

Humans perform and rely on face recognition routinely and effortlessly throughout their daily lives. Multiple works in recent years have sought to replicate this process in a robust and automatic way. However, it is known that the performance of face recognition algorithms is severely compromised in non-ideal image acquisition scenarios. In an attempt to deal with conditions, such as occlusion and heterogeneous illumination, we propose a new approach motivated by the global precedent hypothesis of the human brain's cognitive mechanisms of perception. An automatic modeling of SIFT keypoint descriptors using a Gaussian mixture model (GMM)-based universal background model method is proposed. A decision is, then, made in an innovative hierarchical sense, with holistic information gaining precedence over a more detailed local analysis. The algorithm was tested on the ORL, ARand Extended Yale B Face databases and presented state-of-the-art performance for a variety of experimental setups.

KEYWORDS

biometrics; face recognition; partial data; universal background model; Gaussian mixture models

1. INTRODUCTION

Personal identification plays an important role in almost everyone's daily activities. Knowledge-based and token-based automatic personal identification are the most used techniques to tackle this problem. Token-based approaches take advantage of a personal item to distinguish between individuals, whereas knowledge-based approaches are based on something the user knows to which, theoretically, nobody else has access [1].

Both of these approaches present obvious disadvantages: tokens may be lost, stolen or forgotten, while passwords can easily be forgotten by a valid user or guessed by an unauthorized one [2]. In fact, one can summarize the problem of these approaches by pointing that any piece of material or knowledge can be fraudulently acquired.

Biometrics can be seen as a return to a more natural way of identification. By attempting identification based on physiological and behavioral traits, we are testing someone by who (s)he is, instead of relying on something (s)he owns or knows. Such an approach seems likely to be the way forward [3].

Over the past few years, the issue of face recognition has been in the spotlight of many research works in pattern recognition, due to its wide array of real-world applications. The face is a natural, easily acquirable and usable trait with a high degree of uniqueness, representing one of the main sources of information during human interaction [4]. These marked advantages, however, fall short when images of limited quality, acquired under unconstrained environments, are presented to the system.

It has been noted that the performance of face recognition algorithms is severely compromised when dealing with non-ideal scenarios, such as non-uniform illumination, pose variations, occlusions, expression changes and radical appearance changes [5]. Whereas technological improvements in image capturing and transmitting equipment managed to attenuate most noise factors, partial face occlusions still pose a genuine challenge to automated face recognition [6].

Facial occlusions may occur due to a multiplicity of deliberate or unintentional reasons. Whereas accessories, such as sunglasses and scarves, and facial hair represent quite common sources of occlusion in daily life, they can also be explored by bank robbers and shop thieves in an attempt to avoid recognition. Furthermore, the use of some accessories might be enforced in restricted environments (such as medical masks in hospitals and protection helmets in construction areas) or by religious or cultural constraints [5]. The fact that humans perform and rely on face recognition routinely and effortlessly throughout their daily lives leads to an increased interest in replicating this process in an automated way, even when such limitations are known to frequently occur [7].

Even though there is no consensus in the cognitive science field as to how the human brain recognizes faces, either based on their individual local features or, more holistically, on the basis of their overall shape [8], several works have shown that both levels of information play a non-negligible role in human face perception [9,10]. Whereas holistic representations provide a global summary of the spatial arrangement of contours and textures in an image, local features provide a more detailed regional description of the parts that compose it [11].

In the present work, we propose a robust alternative to face recognition under partial occlusions and variable illumination. An innovative hierarchical decision framework, incorporating both holistic and local descriptions, is proposed. The global precedent hypothesis for human perception [12] is the basis of this new decision strategy. Such a hypothesis claims that face recognition is performed by the human brain in a global-to-local flow, with holistic information gaining precedence over a more detailed local analysis. By following this rationale, we aim to replicate the cognitive process of face recognition by the human brain in an automated way. We evaluate the proposed algorithm on three widely-studied databases—the ORL, ARand the Extended Yale B databases—characterized by a variety of occlusions, small pose variations, facial expressions and illumination conditions.

The rest of the paper is organized as follows: Section 2 summarizes the most recent trends of research in the field of unconstrained face recognition. Section 3 outlines and motivates the proposed algorithm. Section 4 summarizes the most relevant experimental results obtained, as well as a detailed analysis and comparison with the state-of-the-art, and finally, in Section 5, we present the conclusions of the present work, as well as some suggestions for future improvements.

2. RELATED WORK

Face recognition has been a widely-studied research topic in the last few decades. Some traditional approaches, like eigenfaces [13], Fisherfaces [14] and active appearance models [15], have become highly popular and laid out the foundations for a variety of commercial off-the-shelf systems. However, all of the previously mentioned techniques stumble upon the limitations presented in the last section: when non-ideal conditions are present during the acquisition step, recognition performance is severely compromised. The need to improve the state-of-the-art in face recognition to encompass a set of more realistic applications has, thus, been catalyzing research in the area to a set of new directions.

The study of invariant features to diminish the effects of occlusion, illumination and other nefarious sources of noise represents one of the most significant focuses of recent research. Liao *et al.* [16] tried to overcome the need for face alignment that characterizes most holistic approaches employing a multi-keypoint descriptor representation. In this work, any type of face image, either holistic or partial, can be probed for recognition, regardless of the global content. Nallammal and Radha [17] propose a non-negative matrix factorization (NMF) variation to explore the potential of the eyes and the bottom face regions for recognition when the probe images present a high degree of occlusion. An alternative approach is followed by Oh *et al.* [18], where random horizontal and vertical patches of face images are used as templates for cancelable identity verification. Such a technique intentionally distorts biometric information, in a repeatable, but non-reversible manner, to better deal with the compromising of biometric templates. Karande and Talbar [7] address the problem of face recognition with large rotation angles and variable illumination conditions through the use of edge information for independent component analysis (ICA). The work by Geng and Jiang [19] explores some of the known limitations of the widely-studied SIFT approach, for a specialized application in face image description. An alternative keypoint detection and a partial descriptor are both proposed in an attempt to adapt the traditional algorithm to non-rigid and smooth objects. Cho *et al.* [11] propose a two-step approach to face recognition, with principal component analysis (PCA) used at a coarser level and Gabor filtering at a finer level. This finer analysis in only carried out if the coarser recognition results do not present a high degree of reliability. Recently, Facebook's Deep Face Project claimed 97.25% accuracy, where humans achieve an accuracy of 97.53%. Their approach, published by Taigman *et al.* [20], was based on deep neural networks, allowing the effective use of highly complex statistical models trained for large volumes of data.

Recently, approaches based on sparse representation classification (SRC) have shown impressive performance in unconstrained face recognition and became one of the hot research topics in the area. The first reported use of SRC for face recognition, by Wright *et al.* [21], approached the problem of partial occlusion by representing face images as a linear combination of the whole face gallery and a vector of residuals at the pixel level. Classification was then achieved by $l_1$minimization of the vector of residuals for each possible identity. Zhou *et al.* [22] further Improved this methodology by enforcing spatial coherence of occluded pixels through the use of Markov random fields (MRF). The spatial continuity of occlusions in face images was also explored by Qian*et al.* [23]. Their methodology takes advantage of the low-rank error images that are originated in occluded images by traditional SRC methodologies, to perform effective and robust recognition when such conditions are observed. Besides spatial coherence of occlusions, some other topics have been explored in recent works as possible improvements to the original SRC proposal. Wang*et al.* [24] propose an Adaptive SRC (ASRC) approach capable of selecting the most discriminative samples for each representation, using joint information from both sparsity (l_1 minimization), as well as correlation (l_2 minimization). Shen *et al.* [25] propose a variation of SRC for implementation in Android and iOS mobile devices. Their proposal optimizes the computation of residual values with significant gain in computational efficiency and no considerable losses in recognition accuracy. Jian *et al.* [26] also center their attention on the computational speed limitations of the SRC approach. Their proposal, based on the orthogonal matching pursuit (OMP) algorithm, achieves fast and robust face recognition, even though the best results are only achieved through a preliminary occlusion detection block. Even though considerable work has been performed in the area, the main drawback regarding the SRC approach is still posed by the need for an extensive and diverse library of well-aligned face examples.

Another focus of research in recent years concerns the use of prior knowledge regarding occlusions in face images. Zhang *et al.* [27] proposed an estimation of the probability distribution of occlusions in feature space using the Kullback–Leibler divergence (KLD). In a mixed approach regarding both previous detection of occlusions and SRC face recognition, Li *et al.* [6] present a two-step SRC approach, where SRC is both used to first discriminate occluded pixels from unoccluded regions and then to perform face recognition. The use of downsampled images allows a significant improvement in processing speed. Min *et al.* [5], on the other hand, perform occlusion detection using MRF to promote the spatial coherence of the detected occlusion regions. The recognition step is then carried out solely on the non-occluded regions. Even though the *a priori*detection of occlusions may significantly improve the accuracy of local face recognition, the introduction of a new block in the recognition system may bring about a new set of problems. An increase in the computational cost of the process, as well as the creation of a new source for errors that may condition the recognition process from its earlier steps can be counted among such challenges.

In the present work, we propose a robust approach to face recognition when non-ideal conditions, such as partial occlusions and severe illumination variations, affect the acquisition environment of the system. In an attempt to tackle most of the limitations presented in the works outlined in the last paragraphs, we designed a new hierarchical recognition framework. This innovative approach allows a considerable reduction in the computational cost of the whole recognition process, while also allowing an intuitive integration of multiple region-based details. The proposed algorithm is able to achieve accurate face recognition, even when a limited set of images with small variations is used for model training.

2.1. Algorithm Overview

The proposed algorithm is schematically represented in Figure 1. Figure 1a,b depicts the enrollment process in the proposed approach. During enrollment, a new individual's biometric data are inserted into a previously existent database of individuals. In the present work, a hierarchical ensemble of i partial face models is trained for each enrolled subject. The *M*individual-specific models are built by maximum *a posteriori* (MAP) adaptation of the corresponding set of *M* universal background models (UBM) using individual-specific data. The UBM is a representation of the distribution that a biometric trait presents in the universe of all individuals. MAP adaptation works as a specialization of the UBM based on each individual's biometric data. The idea of MAP adaptation of the UBM was first proposed by Reynolds [28], for speaker verification, and will be further motivated in the following sections.

3. PROPOSED METHODOLOGY

The database is probed during the recognition process to assess either the validity of an identity claim (verification) or the *k* most probable identities (identification) given an unknown sample of biometric data. In the present work, we propose an innovative approach to the recognition process based on the global precedence hypothesis of face perception by the human brain. Recognition is performed hierarchically, as depicted in Figure 1c, with global models taking precedence over more detailed ones. Partial models are hierarchically organized into levels. Each level is composed by a set of non-superimposing subregions, I_l, of equal size (Levels 2–3 and 4–5 were hierarchically ordered in an arbitrary order, even though their composing regions are of equal size. Previous knowledge of expected types of occlusion could be explored when specifying this order.). Subregions at the same level sum to the full-face image, I_0. During recognition, a test image from an unknown source follows the hierarchical flow depicted in Figure 1c, until a decision can be made with a significant degree of certainty. The significance of a decision carried out at a single level is defined through the analysis of the likelihood-ratio values obtained for each possible identity claim. Decisions are made independently for each subregion at the same level, and only the most significant one is kept. The following sections will further detail the process of model training and recognition score computation for a single generic subregion, while also motivating the proposed UBM framework.

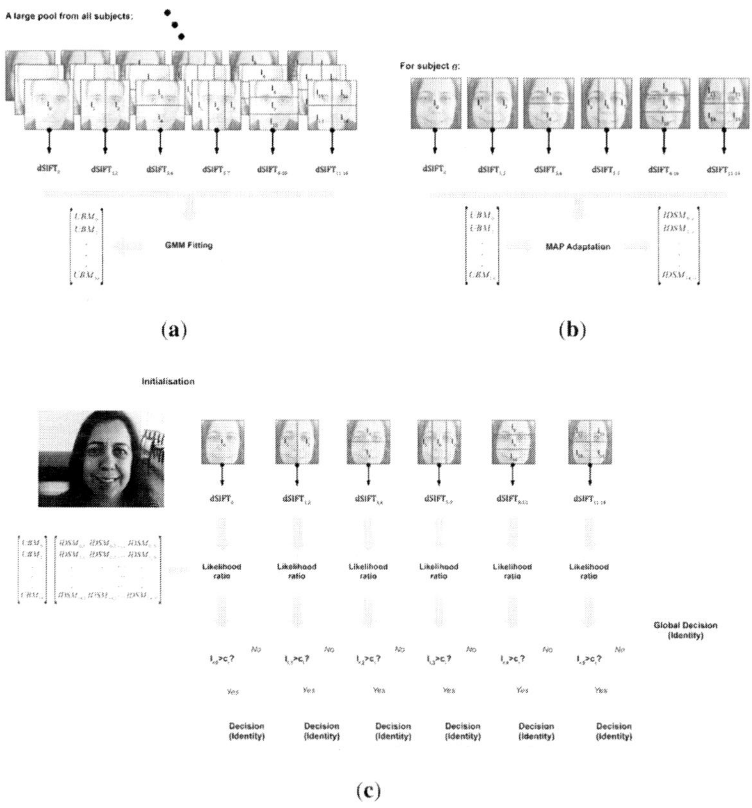

Figure 1. Schematic representation of the proposed algorithm and its main blocks: (**a**) training of the universal background models using data from multiple individuals; (**b**) maximum *a posteriori* (MAP) adaptation of the universal background models (UBM) to generate individual specific models; and (**c**) testing with new data from unknown sources.

3.1. Universal Background Model

The universal background modeling strategy was initially proposed in the field of voice biometrics [29]. Its framework can be easily understood if the problem of biometric verification is interpreted as a basic hypothesis test. Given a biometric sample *Y* and a claimed ID, *S*, we define:

$$H_0\text{: } Y \text{ belongs to } S$$
$$H_1\text{: } Y \text{ does not belong to } S$$

as the null and alternative hypothesis, respectively. The optimal decision is made by a likelihood-ratio test:

$$\frac{p(Y|H_0)}{p(Y|H_1)} \begin{cases} \geq \theta & \text{accept } H_0 \\ < \theta & \text{accept } H_1 \end{cases} \tag{1}$$

where ϑ is the decision threshold for accepting or rejecting H_0 and $p(Y \mid H_i)$ is the likelihood of observing *Y* knowing that H_i is true. The goal of a biometric verification system can, thus, be accomplished by the computation of the likelihood values $p(Y|H_0)$ and $p(Y|H_1)$ for a given sample. It is intuitive to note that H_0 will correspond to a model λ_{hyp} that characterizes the hypothesized individual, whereas H_1 will represent the alternative hypothesis, that is the model of all of the alternatives to the hypothesized individual, $\lambda_{\overline{hyp}}$.

The computation of $p(Y|H_i)$ depends on the specific strategy chosen for data modeling. If H_0 and H_1 correspond to a pair of generative models, trained on sets of genuine and impostor data, respectively, then the $p(Y|H_i)$ values can be roughly expressed as the projections of biometric sample *Y* onto each of these models. This formulation motivates the need for a model that successfully covers the space of alternatives to the hypothesized identity. The most common designation in the literature for such a model is the universal background model or UBM [30]. Such a model must be trained on a rather large set of data, so as to faithfully cover a representative user space.

Even though the UBM approach was initially proposed for verification mode, we extrapolate its rationale for identification systems. Instead of performing a single one *vs.* one likelihood-ratio test and checking the validity of the condition presented in Equation (1), a one *vs.* all approach may alternatively be considered. Given an unknown sample, the most likely identity, Id_{max}, will correspond to the highest likelihood-ratio value, amongst all enrolled users:

$$Id_{max} = \underset{i}{\operatorname{argmax}} \left(\frac{p(Y|H_0^{(i)})}{p(Y|H_1)} \right) \tag{2}$$

where $H_0^{(i)}$ represents the model describing user *i*.

Defining an objective way of quantifying $p(Y|H_0)$ and $p(Y|H_1)$ becomes, thus, the true challenge when following this approach. In the following sections, we analyze in detail the strategies chosen to model both λ_{hyp} and $\lambda_{\overline{hyp}}$.

3.2. Hypothesis Modeling

In the present work, we chose Gaussian mixture models (GMM) to model both the UBM, *i.e.*, $\lambda_{\overline{hyp}}$, and the individual-specific models (IDSM), *i.e.*, λ_{hyp}. Such models are capable of capturing the empirical probability density function of a given set of feature vectors, so as to faithfully model their intrinsic statistical properties [28]. The choice of GMM to model feature distributions in biometric data is extensively motivated in many works of related areas. From the most common interpretations, GMMs are seen as capable of representing broad "hidden" classes, reflective of the unique structural arrangements observed in the analyzed biometric traits [28]. Besides this assumption, Gaussian mixtures display both the robustness of parametric unimodal Gaussian density estimates, as well as the ability of non-parametric models to fit non-Gaussian data [31]. This duality, alongside the fact that GMM have the noteworthy strength of generating smooth parametric densities, confers such models a strong advantage as generative models of choice. For computational efficiency, GMM models are often trained using diagonal covariance matrices. This approximation is often found in the biometrics literature, with no significant accuracy loss associated [32].

All models are trained on densely-sampled sets of scale-invariant feature transform (dSIFT) keypoint descriptors, extracted from previously normalized facial sub-regions. Illumination normalization is performed using the Weber-face approach [33]. Even though traditional SIFT descriptors present invariance to a set of common undesirable factors (image scaling, translation, rotation), thus conferring them a strong appeal in unconstrained biometrics, the fact that they fail to adapt to heterogeneous illumination conditions severely hinders their practical use in real-life applications. The normalization step is, therefore, of the utmost importance.

Dense SIFT is a variation of the traditional SIFT methodology [34], where keypoint descriptors are extracted in a roughly equivalent manner to running SIFT on a dense grid of locations at a fixed scale and orientation [35]. Dense sampling mitigates the potential errors introduced by the detection of an unreliable set of interest points in its sparse counterpart [36]. In the present work, we train GMMs on the set of all densely-sampled keypoint descriptors from all individuals (UBM) and adapt individual models using data from specific subjects alone (IDSM), achieving a stable summary of the image content for every enrolled user.

Originally, SIFT descriptors were defined in 128 dimensions. However, we chose to perform a PCA, as suggested in [37], reducing the dimensionality to 32. Such a reduction allows not only a significant reduction in the computational complexity of the training phase, but also an improved distinctiveness and robustness to the extracted feature vectors, especially as far as image deformation is concerned [38]. We computed the principle components from the same set of keypoint descriptors used to train the UBM.

3.2.1. UBM Parameter Estimation

To train the universal background model, a large amount of "impostor" data, *i.e.*, a set composed of data from all the enrolled individuals, is used, so as to cover a wide range of possibilities in the individual search space [37]. The training process of the UBM is simply performed by fitting a *k*-mixture GMM to the set of PCA-reduced feature vectors extracted from all of the "impostors".

If we interpret the UBM as an "impostor" model, its "genuine" counterpart can be obtained by adaptation of the UBM's parameters, $\lambda_{\overline{hyp}}$. using individual specific data. For each enrolled user, *n*, an IDSM, defined by parameters. λ_{hyp}, is therefore obtained. The adaptation process will be outlined in the following section.

3.2.2. MAP Adaptation

IDSMs are generated by the tuning of the UBM parameters, in a maximum *a posteriori* (MAP) sense, using individual-specific biometric data. This approach provides a tight coupling between the IDSM and the UBM, resulting in better performance and faster scoring than uncoupled methods, as well as a robust and precise parameter estimation, even when only a small amount of data is available [37]. The adaptation process consists of two main estimation steps. First, for each component of the UBM, a set of sufficient statistics is computed from a set of *M* individual-specific feature vectors, $X = \{\mathbf{x}_1, \ldots, \mathbf{x}_M\}$:

$$n_i = \sum_{m=1}^{M} p(i|\mathbf{x}_m) \tag{3}$$

$$E_i(\mathbf{x}) = \frac{1}{n_i}\sum_{m=1}^{M} p(i|\mathbf{x}_m)\mathbf{x}_m \quad (4)$$

$$E_i(\mathbf{x}\mathbf{x}^t) = \frac{1}{n_i}\sum_{m=1}^{M} p(i|\mathbf{x}_m)\mathbf{x}_m\mathbf{x}_m^t \quad (5)$$

where $p(i|\mathbf{x}_m)$ represents the probabilistic alignment of $\mathbf{x}_m$ into each UBM component. Each UBM component is then adapted using the newly-computed sufficient statistics and considering diagonal covariance matrices. The update process can be formally expressed as:

$$\hat{w}_i = [\alpha_i n_i/M + (1-\alpha_i)w_i]\,\xi \quad (6)$$

$$\hat{\boldsymbol{\mu}}_i = \alpha_i E_i(\mathbf{x}) + (1-\alpha_i)\boldsymbol{\mu}_i \quad (7)$$

$$\hat{\Sigma}_i = \alpha_i E_i(\mathbf{x}\mathbf{x}^t) + (1-\alpha_i)(\boldsymbol{\sigma}_i\boldsymbol{\sigma}_i^t + \boldsymbol{\mu}_i\boldsymbol{\mu}_i^t) - \hat{\boldsymbol{\mu}}_i\hat{\boldsymbol{\mu}}_i^t \quad (8)$$

$$\boldsymbol{\sigma}_i = \mathrm{diag}(\Sigma_i) \quad (9)$$

where $\{w_i, \boldsymbol{\mu}_i, \boldsymbol{\sigma}_i\}$ are the original UBM parameters and $\{\hat{w}_i, \hat{\boldsymbol{\mu}}_i, \hat{\boldsymbol{\sigma}}_i\}$ represent their adaptation to a specific speaker. To assure that $\sum_i w_i = 1$, a weighting parameter ξ is introduced. The α parameter is a data-dependent adaptation coefficient. Formally, it can be defined as:

$$\alpha_i = \frac{n_i}{r + n_i} \quad (10)$$

The relevance factor r weights the relative importance of the original values and the new sufficient statistics. In the present, work we set $r = 16$.

3.3. Hierarchical Decision

The whole training process is repeated at most M times for each of M facial subregions, defined *a priori*. In the present work, we used the $M = 14$ regions previously exemplified in Figure 1.

Traditionally, the recognition phase with new data from an unknown source is a fairly simple process. The new test data, $X_{test} = \{\mathbf{x}_{t,1},\ldots, \mathbf{x}_{t,N}\}$, where $\mathbf{x}_{t,i}$ is the i-th PCA-reduced SIFT vector extracted from a given subregion m of test subject t, is projected onto both the UBM and either the claimed IDSM (in verification mode) or all such models (in identification mode). The recognition score, $s_{t,m}$, is obtained as the average likelihood-ratio of all keypoint descriptors $\mathbf{x}_{t,i}$, $s_{t,m} = \frac{1}{N}\sum_{i=1}^{N} s_{t,m}^{(i)}$. The decision is then carried out by checking the condition presented inEquation (1), in the case of verification, or by detecting the maximum likelihood-ratio value for all enrolled IDs (Equation (2)), in the case of identification.

Such a decision step represents the second big advantage of the UBM approach. The ratio between the IDSM and the UBM probabilities of the observed data is a more robust decision criterion than relying solely on the IDSM probability. This results from the fact that some subjects are more prone to generate high likelihood values than others, *i.e.*, some people have a more "generic" look than others. The use of a likelihood ratio with a universal reference works as a normalization step, mapping the likelihood values in accordance with their global projection. Without such a step, finding a global optimal value for the decision threshold, ϑ, presented inEquation (1), would be a far more complex process.

In an attempt to integrate meaningful information from the *M* facial subregions and to deal better with partial or missing data situations originated by occlusions, an hierarchical recognition framework is proposed. The rationale behind the methodology described below can be easily understood if some studies of the human brain's cognitive mechanisms of perception are taken into consideration. One such work, proposed in 1977 by David Navon [12], describes the aforementioned mechanism as a hierarchical process, where holistic representations precede more detailed local features. If we interpret this perception mechanism in the scope of human face recognition, we can conclude that an attempt to classify the face at a global level is the starting point to the whole recognition process, whereas increasingly detailed descriptions are only taken subsequently if necessary. This conceptual description of the human perception mechanism serves as the basis for the proposed hierarchical recognition algorithm, whose flowchart is presented inFigure 1c.

The main steps of the proposed hierarchical identification algorithm are as described below:

1 Initialization: Starting with the full-face image from an unknown user, I_0, a densely-sampled grid of SIFT keypoint descriptors is extracted; the likelihood values, $l_{t,0}^{IDSM_t}$ and $l_{t,0}^{UBM_0}$, are then computed for the IDSM of every enrolled user $t \in \{1..T\}$ and the UBM of the tested region; the recognition scores for every possible identity, $\mathbf{s}_{t,0} = \{s_{1,0}, \ldots, s_{T,0}\}$, are then computed through Equation (1).

2 Certainty index computation: The certainty index of a given region, c_m, measures how likely it is that the obtained vector of likelihood ratios, $\boldsymbol{s}_{t,m}$, corresponds with the ideal case of a single correct identity match. If no false positives corrupt the vector of likelihood ratios obtained for a single image, a significant difference will be observed between the highest value, $\boldsymbol{s}_{t^*,m}$, (true identity) and the average of all other values, $\frac{1}{T-1}\sum_{t=1,t\neq t^*}^{T} \boldsymbol{s}_{t,m}$, (average impostor). The certainty index can thus be interpreted as a degree of separability between these two quantities:

$$c_m = \boldsymbol{s}_{t^*,m} - \frac{1}{T-1}\sum_{t=1,t\neq t^*}^{T} \boldsymbol{s}_{t,m} \tag{11}$$

3 Decision to go to next level: If the c_m value exceeds a previously optimized threshold, ϑ_l, the maximum likelihood-ratio decision is accepted. When $c_m < \vartheta_l$, however, the algorithm will consider that an analysis at a more detailed level is necessary to achieve a decision with a higher degree of confidence. At this point, the algorithm proceeds to the next level, working on subregions l_{1-2}, the second in the hierarchical chain depicted in Figure 1. When one level is composed by multiple subregions, each one of them is treated independently, and only the maximum c_m value among them is considered for the decision criterion:

$$c_m = \begin{cases} c_0 & \text{if } l = 1 \\ \max\{c_1, c_2\} & \text{if } l = 2 \\ \max\{c_3, c_4\} & \text{if } l = 3 \\ \max\{c_5, c_6, c_7\} & \text{if } l = 4 \\ \max\{c_8, c_9, c_{10}\} & \text{if } l = 5 \\ \max\{c_{11}, c_{12}, c_{13}, c_{14}\} & \text{if } l = 6 \end{cases} \tag{13}$$

4 Repeat: Steps 1 to 3 are hierarchically repeated for every level until $c_m > \vartheta_l$. If all L levels are considered and none is able to achieve a significant decision, the decision corresponding to the highest value of c_m, $\max_m(c_m)$, amongst all L levels is considered.

As the proposed algorithm is capable of performing recognition without the need of processing all subregions, computational speed is significantly improved over simpler approaches that explore the fusion of all local recognition scores. Furthermore, the proposed algorithm is capable of automatically deciding if a more detailed exploration of local features is necessary or if the information obtained up to a certain point is enough to make a decision. This autonomy alongside the intuitive notion of the global precedence hypothesis are the most notorious strengths of the proposed recognition algorithm.

An alternative to deciding non-classified images at the end of the hierarchical chain was also considered. In this new setting, if no level is capable of making a decision according to the aforementioned criteria, images are kept as "doubtful", and no decision is made, in a process similar to classification with a reject option [39]. This approach may be thought of as a viable alternative for real-life applications, where feedback to the user can be explored to adjust the environmental conditions in severely unconstrained scenarios.

4. RESULTS AND DISCUSSION

4.1. Experimental Setups

The proposed algorithm was tested on the Database of Faces (formerly "The ORL Database of Faces"), the Extended Yale Face Database B and the AR face database. All tested databases are widely known for their diversity of pose, illumination and occlusion conditions, respectively. The next sections outline the main features of each database, as well as the experimental setup chosen for training and testing with each of them in the scope of the present work.

4.1.1. ORL

The Database of Faces (formerly "The ORL Database of Faces") [40] contains 400 images from 40 subjects, divided equitably with a total of 10 images per individual. Images were taken at different points in time with variable lighting, expression and pose conditions. For performance assessment, we use a single sample from each individual for training and the remaining nine images for testing. This process is repeated for each possible training image, and the performance is computed as the average of the 10 runs. An alternative approach is also explored by using multiple templates per subject instead. In this new setup, the first five samples per subject are used for training, while the remaining five samples are used for testing. Example images from two subjects may be observed in Figure 2.

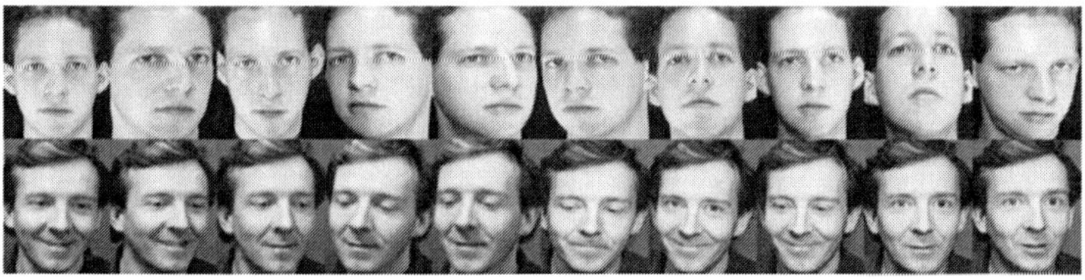

Figure 2. Example images from two subjects of the ORLdatabase.

4.1.2. Extended Yale B

The Extended Yale Face Database B [41] is composed of 2432 images corresponding to a total of 38 individuals. All images are frontal faces acquired under varying illumination conditions. The database is divided into five subsets, numbered 1 to 5, according to the ranges of angles between the light source direction and the camera axis. An example of all images from a single subject may be observed in Figure 3.

All images from Subset 1 were used for model training, while all other subsets were tested independently, so as to better assess the robustness of the proposed algorithm to a variety of illumination conditions.

4.1.3. AR

The AR database [42] contains over 4000 frontal face images from 126 individuals, acquired under variable illumination, expression and occlusion. Occlusions can be divided into two main categories: sunglasses and scarf. An example of all images from a single individual is presented inFigure 4.

All unoccluded images from every individual are chosen to train both the local UBMs and IDSMs. The remaining scarf and sunglasses occluded images are tested separately, so as to better analyze the consistency of the proposed algorithm when exposed to variable types of occlusion.

4.1.4. Performance Analysis

Figures 5, 6 and 7 depict the most relevant results obtained by the proposed algorithm for the ORL, Extended Yale B and AR databases, using the previously mentioned training and testing setups. Furthermore, Tables 1, 2 and 3 present a comparison between the proposed work and the reported performance in some recent works, performed under similar experimental conditions. We chose to assess the rate of correctly identified individuals, by checking if the true identity is present among the *N* highest ranked identities. The *N* parameter is generally referred to as rank. This allows us to define the Rank-1 recognition rate, r_1, as the recognition rate at $N = 1$.

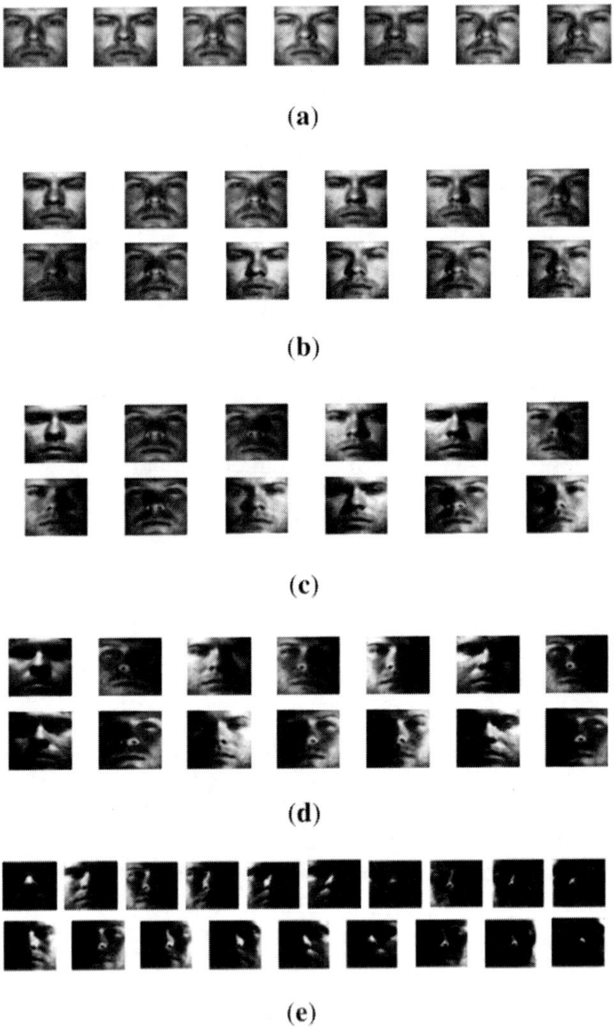

Figure 3. All images from a single subject enrolled in the Extended Yale B database. Images (**a**) to (**e**) correspond to Subsets 1 to 5, respectively.

Figure 4. Example images from one subject of the ARdatabase.

Each plotted point refers to a single ϑ_l value, ranging from [0, ∞]. For each tested ϑ_l value, we plotted a series of performance metrics against the corresponding average processing time per image. Using the ϑ_l = 0.1 vertical line from Figure 7a as a reference, we can distinguish three metrics that are common to all tested setups. The black line represents the evolution of r_1 when "doubtful" images are classified through the aforementioned criterion, maxm(cm). When no such *a posteriori* classification is performed, a set of images is left unclassified, as no level from the hierarchical chain was shown to present enough detail to perform an accurate recognition. The red line represents the evolution of the "doubtful" image percentage with increasing ϑ_l values. It is intuitive to note that lower ϑ_l values generate lower reject-option classification ratios, as a smaller dissimilarity between individuals will still trigger a "certain" decision. If no "doubtful" images are considered in the performance evaluation of the proposed algorithm, then the r_1 computation is performed only with respect to the non-rejected images. The blue line represents this alternative evaluation, and the corresponding reject percentage can be easily traced to the equivalent ϑ_l point in the red line. Taking, once again, the ϑ_l = 0.1 vertical line from Figure 7a as a reference, we can see that the proposed methodology yields a r_1 value of roughly 98.00%. If the approximately 30% "doubtful" images are not considered for evaluation, the r_1 value increases to 100.00%, with respect to the non-rejected images.

For a better comprehension and deeper analysis of the method, we chose to assess performance in three specific points:

- Optimal $\boldsymbol{\vartheta_l}$ value: We consider the optimal ϑ_l value as the point where a visible performance plateau is achieved in the time *vs.* performance plot. For the *AR* database, this value was set to ϑ_l = 0.4, whereas for the Extended Yale B Face database, it was set to ϑ_l = 0.15. For the ORL database, the ϑ_l value was optimized for each of the aforementioned experimental setups. For the single template approach, the value was set to ϑ_l = 0.02, while for multiple templates, it was set to ϑ_l = 0.15.

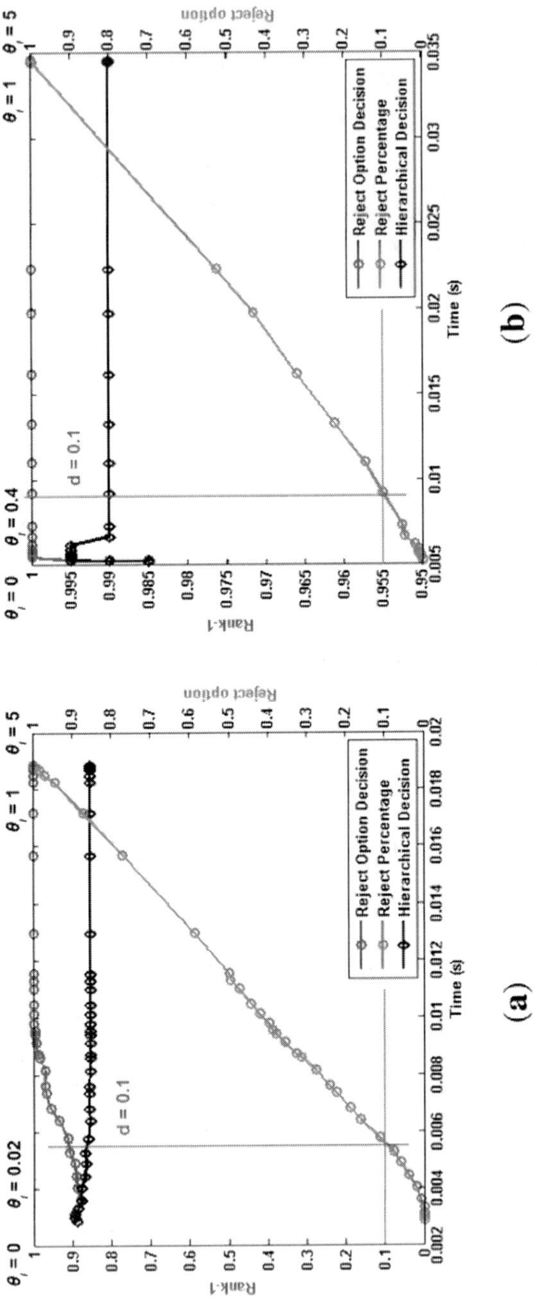

Figure 5. Main results obtained with the proposed methodology for the sunglasses and scarf images of the ORL face database. Each plotted point represent a specific value of parameter ϑ_l, ranging from [0, ∞].

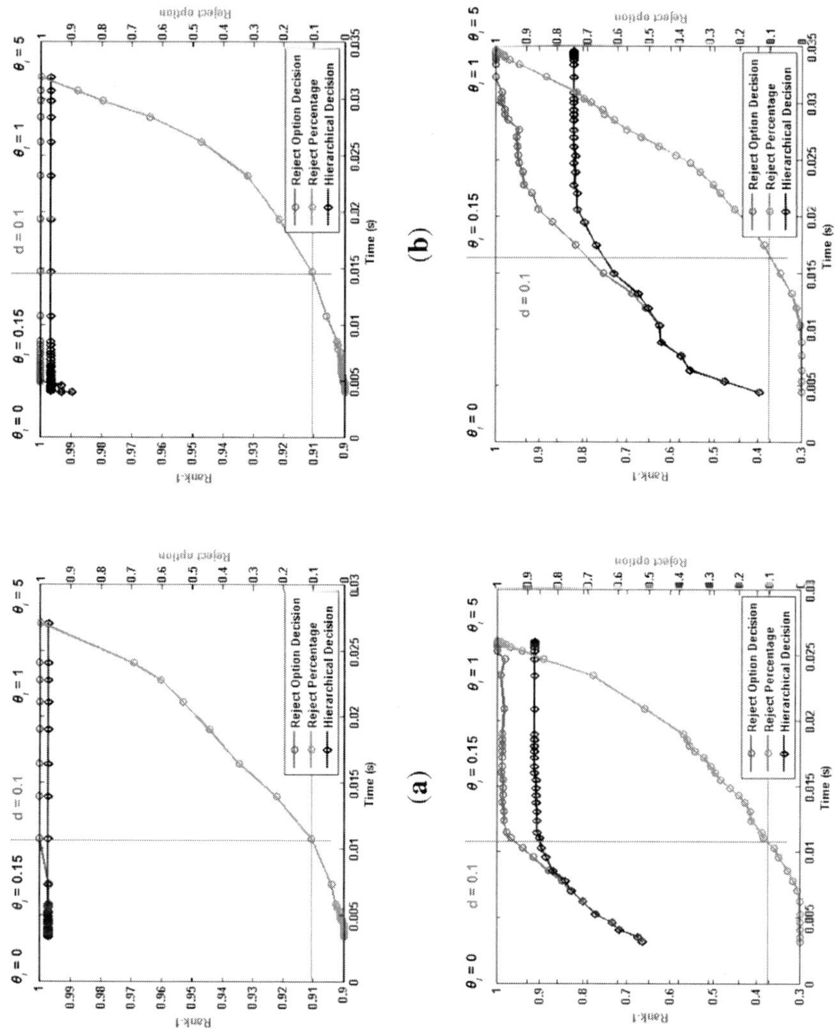

Figure 6. Main results obtained with the proposed methodology for Subsets 2–5 of the Extended Yale B Face database.

- $\vartheta_l \rightarrow \infty$: extreme behavior when the ϑ_l parameter is set to high values.

- $d < 0.1$: point where the ratio of non-classified images at the end of the hierarchical chain reaches 10% of all tested images.

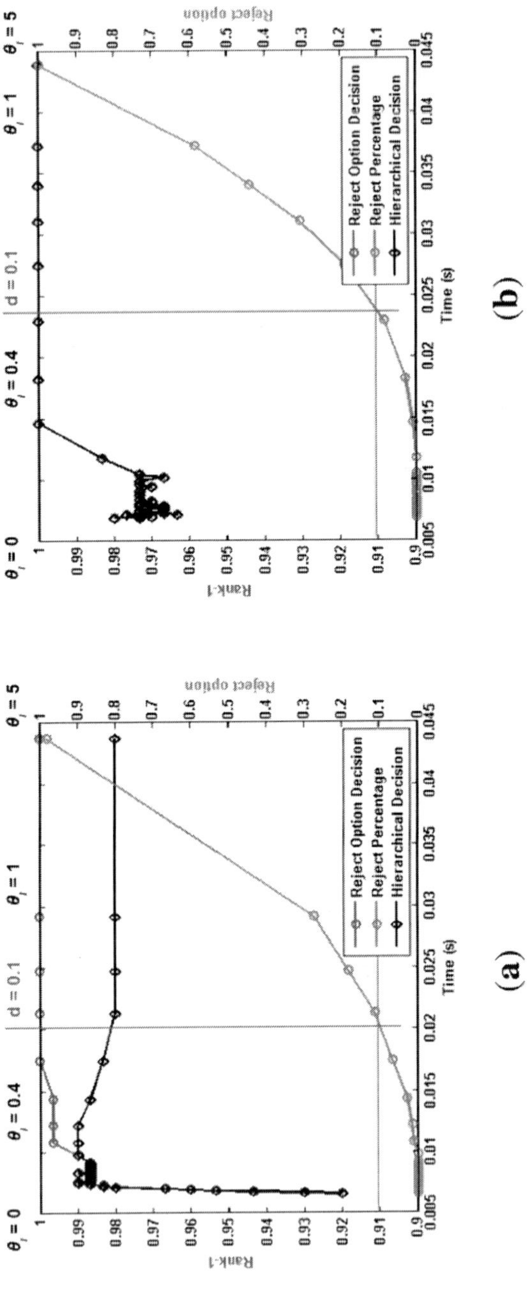

Figure 7. Main results obtained with the proposed methodology for the sunglasses and scarf images of the AR face database. Each plotted point represent a specific value of parameter θ_l, ranging from [0, ∞].

Table 1. Comparison between the obtained r_1 values for the ORL database and some state-of-the-art algorithms.

Work	Single Template $\mu \pm \sigma$	Multiple Templates
Proposed$_{\theta=0.02\|0.15}$	**89.45% ± 1.24%**	**99.5%**
Proposed$_{\theta\to\infty}$	85.79% ± 1.85%	99.00%
Proposed$_{d<0.1}$	***93.60% ± 1.85%***	***100.0%***
Geng and Jiang [?]	89.0%	**99.5%**
Zhang *et al.* [?]	–	85.0%
Xiao *et al.* [?]	–	97.50
Xu *et al.* [?]	–	97.10

Table 2. Comparison between the obtained r_1 values for the AR database and some state-of-the-art algorithms.

Work	AR Sunglasses	AR Scarf	AR Average
Proposed$_{\theta=0.4}$	**100.0%**	99.00%	**99.50%**
Proposed$_{\theta\to\infty}$	**100.0%**	98.00%	99.00%
Proposed$_{d<0.1}$	***100.0%***	***100.0%***	***100.0%***
Min *et al.* [?]	75.00%	92.08%	83.54%
Li *et al.* [?]	97.00%	52.00%	74.50%
Morelli Andrés *et al.* [?]	97.46%	99.15%	98.31%
Qian *et al.* [?]	99.00%	**100.0%**	**99.50%**

A thorough analysis of Figures 5, 6 and 7 leads to some interesting conclusions regarding the behavior of the proposed algorithm under variable image acquisition conditions. The performance metric r_1 was computed for a set of ϑ_l values, ranging from [0, ∞]. It is trivial to understand that lower values of ϑ_l will lead the proposed algorithm to an extreme case where all images are classified in the first level of information, *i.e.*, the full face, thus reducing the computational complexity and average processing time. On the other hand, when $\vartheta \to \infty$, all images will reach the end of the hierarchical chain without a certain decision having been made. In this opposite extreme behavior, all images are classified *a posteriori* using information from every level, with a significant increase in both computational complexity and performance. With such a wide variety of possibilities, we chose to analyze the global behavior of the proposed work by plotting the evolution of r_1 values against the average processing time, when the ϑ_l parameter goes from [0, ∞].

Table 3. Comparison between the obtained r_1 values for the Extended Yale B database and some state-of-the-art algorithms.

Work	Yale B SS_2	Yale B SS_3	Yale B SS_4	Yale B SS_5
Proposed$_{\theta=0.15}$	99.71%	99.66%	90.91%	81.99%
Proposed$_{\theta\to\infty}$	99.71%	99.66%	91.20%	81.99%
Proposed$_{d<0.1}$	*100.0%*	*100.0%*	*96.68%*	*81.66%*
Cho *et al.* [?]	100%	100%	99.2%	—
Yang *et al.* [?]	—	—	90.2%	47.9%
Jiang *et al.* [?]	100%	99.12%	83.46%	57.84%
Jia *et al.* [?]	100%	100%	**97.15%**	**96.13%**

Regarding the evolution of r_1 for variable ϑ_l values, we might point out the considerable drop in performance under less ideal conditions, when $\vartheta \to 0$. Whereas Subsets 2 and 3 from the Extended Yale B database show excellent performance from both aforementioned approaches, rivaling those obtained for $\vartheta \to \infty$, such behavior is highly compromised for the more challenging scenarios of Subsets 4 and 5. In such cases, the observed drop in performance is less significant for higher ϑ_l values. Furthermore, the results obtained for the AR database stress that the proposed methodology is capable of consistently presenting high performance regardless of the acquisition conditions and noise factors. Whereas the Extended Yale B database presented the challenge of heterogeneous illumination, which might be understood as a "natural" source of occlusion, the AR database presents the challenge of spatially coherent occlusion regions. For both cases, performance observed for the proposed hierarchical methodology rivals the state-of-the-art. Regarding the $\vartheta \to \infty$ case, the expectation that the best results would be consistently observed for this extreme scenario was fulfilled, at the expense of higher computation complexity and average processing time. Nevertheless, if no time constraints exist in the application scenario in which the proposed methodology is implemented, higher values of ϑ_l seem the ideal choice. The ORL database maintains a somewhat stable performance regardless of the chosen value for the ϑ_l parameter. As the proposed hierarchical methodology is contextually motivated for occlusion scenarios and no such conditions are present in this database, the observed behavior fits the expectations. Regardless of the absence of occlusions in the ORL images, the reject option classifier still proves a useful tool at discriminating "doubtful" images and increasing the reliability of the decisions obtained for the "non-doubtful" ones.

Tables 1, 2 and 3 present a comparative analysis between the proposed methodologies and some state-of-the-art works, in similar experimental setups, for the AR and Extended Yale B databases, respectively. It is readily observed that the proposed algorithm presents the most consistent and robust behavior, regardless of the nature of the present occlusions, for the AR face database. While the work by Li *et al.* [6] presents higher performance for sunglasses occlusion, their performance regarding the alternative scarf occlusion is considerably lower than the one obtained with the proposed algorithm. Min *et al.* [5] present a work that suffers from exactly the opposite problem, with good performance observed for scarf occlusion, but lower results for sunglasses. The aforementioned works seem to suffer from overfitting to certain classes of images and lack the robustness to adapt to new cases. Such robustness to the nature and location of the occlusion can be observed for both the proposed methodology, as well as the works from Morelli *et al.* [46] and Qian *et al.* [23]. Both of these works show a similar trend to the proposed algorithm and high performance for the whole database. Alongside our work, and to the best of our knowledge, they represent the state-of-the-art performance for the AR database in unconstrained face recognition. A similar analysis can be made for both setups under which experiments were carried out with the ORL database. The best performance obtained with the proposed algorithm is on par with the best value found in the literature, in the work by Geng and Jiang [19]. This observation leads to the conclusion that the proposed methodology is capable of achieving state-of-the-art performance for a wider variety of scenarios besides occlusion. The aforementioned observation that performance seems to be independent of the chosen value for the ϑ_l parameter may, however, indicate that a simpler approach than the whole hierarchical chain might also output good results. Using simply the first level, *i.e.*, the full-face images, the r_1 values observed for the single and multiple template setups were 88.10% ± 1.05% and 98.5%, respectively. These results differ very little from the optimal results presented in Table 1, thus corroborating the expectation of good performance for a less computationally complex approach.

Regarding the Extended Yale B Face database, our algorithm is shown to perform similar to state-of-the-art performance for all subsets, except Subset 5. Even for this subset, the obtained performance is still higher than a few recently published works, even though no direct comparison can be performed with the work by Cho *et al.* [11], whose authors claim that its images are "excluded from the experiment because the illumination conditions are so severe that some images are difficult to recognize even to the naked eye".

Observing the behavior of the blue and red lines for high threshold values, we note that a significant amount of images in this subset exceeds the recognition capability of the proposed methodology. The fact that the blue and black lines diverge greatly, when compared to the results observed in other subsets, shows that posterior classification by the maximum c_m value may not be the most beneficial approach when working with such severely degraded images. In such cases, the reject option classification yields considerably higher performance than the posterior classification of doubtful cases, even for low rejection ratios. It is easily noted how the blue and black lines start to diverge for a value of $d \ll 0.1$, as a result of this behavior.

4.2. Implementation Details

The proposed algorithm was developed in MATLAB R2012a and tested on a PC with 3.40 GHz Intel(R) Core(TM) i7-2600 processor and 8 GB RAM. To train the GMMs, we used the Netlab toolbox [50], whereas dSIFT keypoint description was performed using the VLFeat toolbox [51].

5. CONCLUSIONS AND FUTURE WORK

In the present work, we propose an algorithm for face recognition under unconstrained settings, such as heterogeneous illumination and severe occlusions. By training models that only describe features of limited regions of the face, we confer our algorithm a robustness to events where all other regions are occluded. In an attempt to replicate the global precedent hypothesis of the human brain's cognitive mechanisms of perception, we designed an innovative hierarchical recognition algorithm, where face recognition is performed locally only if a more global representation is not capable of achieving a decision with a high degree of certainty. Even though good performance was observed for a wide variety of non-ideal conditions, some ideas are worth noting for future research in the area.

First of all, the choice of SIFT keypoint descriptors distributed in a dense grid is based on a series of assumptions, like constant scale and orientation, that might not always hold. Besides these assumptions, we performed no comparative analysis with other descriptors that might prove more appropriate for face description in less controlled acquisition scenarios. The feature extraction algorithm proposed by Miao and Jiang [52], for example, is reported to be more efficient at extracting interest points from human face images than the SIFT approach followed in the present work and might prove to be an interesting alternative. Still, regarding face representation, no alternative pre-processing methods were considered besides the aforementioned Weber-faces approach. Given that the most significant performance drop of the proposed algorithm was observed for images acquired under significantly low illumination, alternative normalization techniques might deserve further research. A different approach might be considered, as previously mentioned, by introducing a feedback mechanism to the recognition system, with the user being notified when the reject option is triggered in the end of the hierarchical chain. This could allow a user to actively adapt the environmental conditions so as to facilitate a more certain decision.

The choice of the fittest ϑ_l value for each specific application might also motivate a more detailed study. The hypothesis that some users are easier to identify than others inside a given population, an effect known as the Doddington zoo effect [53], suggests that parameter optimization in biometric applications might benefit from a more individual-specific approach. Besides this observation, we fixed the ϑ_l value for all levels of the hierarchical chain. We assumed that due to the normalization of the likelihood-ratio provided by the UBM, recognition scores at different levels share a similar range of values. Even though this assumption seems valid enough in the proposed framework, analyzing the effect of level-specific threshold values might bring about slight increases in performance.

In a final consideration, the proposed work was carried out fully on frontal face images. Most real-life applications in unconstrained scenarios should not enforce such a rigid constraint, and thus, the proposed algorithm should be capable of coping with pose variations. Two alternatives to approach this problem could be the training of multiple models describing individual poses or the training of a single model built on images acquired under multiple poses. Further research into both alternatives would be needed before either of them can be considered as the most fit to expand the proposed framework.

ACKNOWLEDGMENTS

The first author would like to thank Fundação para a Ciência e Tecnologia (FCT), Portugal, for the financial support for the PhD grant with Reference SFRH/BD/87392/2012.

AUTHOR CONTRIBUTIONS

João C. Monteiro and Jaime S. Cardoso conceived and designed the experiments; João C. Monteiro performed the experiments; João C. Monteiro and Jaime S. Cardoso analyzed the data; João C. Monteiro wrote the paper under the guidance of Jaime S. Cardoso.

CONFLICTS OF INTEREST

The authors declare no conflict of interest.

REFERENCES

1. Prabhakar, S.; Pankanti, S.; Jain, A.K. Biometric recognition: Security and privacy concerns.*IEEE Secur. Priv.* **2003**, *1*, 33–42.

2. Jain, A.; Hong, L.; Pankanti, S. Biometric identification. *Commun. ACM* **2000**, *43*, 90–98.

3. Monteiro, J.C.; Sequeira, A.F.; Oliveira, H.P.; Cardoso, J.S. Robust iris localisation in challenging scenarios. *Commun. Comput. Inform. Sci.* **2014**, *458*, 146–162.

4. Castrillón, M.; Déniz, O.; Hernández, D.; Lorenzo, J. A comparison of face and facial feature detectors based on the Viola–Jones general object detection framework. *Mach. Vis. Appl.***2011**, *22*, 481–494.

5. Min, R.; Hadid, A.; Dugelay, J.L. Efficient detection of occlusion prior to robust face recognition. *Sci. World J.* **2014**, *2014*.

6. Li, Y.; Meng, L.; Feng, J.; Wu, J. Downsampling sparse representation and discriminant information aided occluded face recognition. *Sci. China Inf. Sci.* **2014**, *57*, 1–8.

7. Karande, K.J.; Talbar, S.N. *Independent Component Analysis of Edge Information for Face Recognition*; Springer: New Delhi, India, 2014.

8. Tanaka, J.W.; Farah, M.J. Parts and wholes in face recognition. *Q. J. Exp. Psychol.* **1993**, *46*, 225–245.

9. Schwaninger, A.; Schumacher, S.; Bülthoff, H.; Wallraven, C. Using 3D computer graphics for perception: The role of local and global information in face processing. Proceedings of the 4th Symposium on Applied Perception in Graphics and Visualization, Türbingen, Germany, 25–28 July 2007; pp. 19–26.

10. Gold, J.M.; Mundy, P.J.; Tjan, B.S. The perception of a face is no more than the sum of its parts. *Psychol. Sci.* **2012**, *23*, 427–434.

11. Cho, H.; Roberts, R.; Jung, B.; Choi, O.; Moon, S. An Efficient Hybrid Face Recognition Algorithm Using PCA and GABOR Wavelets. *Int. J. Adv. Robot. Syst.* **2014**, *11*, 1–8.

12. Navon, D. Forest before trees: The precedence of global features in visual perception. *Cogn. Psychol.* **1977**, *9*, 353–383.

13. Turk, M.; Pentland, A. Eigenfaces for recognition. *J. Cogn. Neurosci.* **1991**, *3*, 71–86.

14. Belhumeur, P.N.; Hespanha, J.P.; Kriegman, D. Eigenfaces vs. fisherfaces: Recognition using class specific linear projection. *IEEE Trans. Pattern Anal. Mach. Intell.* **1997**, *19*, 711–720.

15. Cootes, T.F.; Edwards, G.J.; Taylor, C.J. Active appearance models. *Lect. Notes Comput. Sci.* **1998**, *1407*, 484–498.

16. Liao, S.; Jain, A.K. Partial face recognition: An alignment free approach. Proceedings of the 2011 International Joint Conference on Biometrics (IJCB), Washington, DC, USA, 11–13 October 2011; pp. 1–8.

17. Nallammal, N.; Radha, V. Comparative Analysis of Partial Occlusion Using Face Recognition Techniques. *Int. J. Image Proc. (IJIP)* **2013**, *7*, 132–139.

18. Oh, B.S.; Toh, K.A.; Choi, K.; Beng Jin Teoh, A.; Kim, J. Extraction and fusion of partial face features for cancelable identity verification. *Pattern Recognit.* **2012**, *45*, 3288–3303.

19. Geng, C.; Jiang, X. Face recognition based on the multi-scale local image structures. *Pattern Recognit.* **2011**, *44*, 2565–2575.

20. Taigman, Y.; Yang, M.; Ranzato, M.; Wolf, L. Deepface: Closing the gap to human-level performance in face verification. Proceedings of the 2014 IEEE Conference on Computer Vision and Pattern Recognition (CVPR), Colombus, OH, USA, 24–27 June 2014; pp. 1701–1708.

21. Wright, J.; Yang, A.Y.; Ganesh, A.; Sastry, S.S.; Ma, Y. Robust face recognition via sparse representation. *IEEE Trans. Pattern Anal. Mach. Intell.* **2009**, *31*, 210–227.

22. Zhou, Z.; Wagner, A.; Mobahi, H.; Wright, J.; Ma, Y. Face recognition with contiguous occlusion using markov random fields. Proceedings of the 2009 IEEE 12th International Conference on Computer Vision, Kyoto, Japan, 27 September– 4 October 2009; pp. 1050–1057.

23. Qian, J.; Yang, J.; Zhang, F.; Lin, Z. Robust Low-Rank Regularized Regression for Face Recognition with Occlusion. Proceedings of the IEEE Conference on Computer Vision and Pattern Recognition Workshops, Colombus, OH, USA, 23–28 June 2014; pp. 21–26.

24. Wang, J.; Lu, C.; Wang, M.; Li, P.; Yan, S.; Hu, X. Robust Face Recognition via Adaptive Sparse Representation. *IEEE Trans. Cybern.* **2014**, *44*, 2368–2378.

25. Shen, Y.; Hu, W.; Yang, M.; Wei, B.; Lucey, S.; Chou, C.T. Face recognition on smartphones via optimised sparse representation classification. Proceedings of the 13th International Symposium on Information Processing in Sensor Networks, Berlin, Germany, 15–17 April 2014; pp. 237–248.

26. Jian, Z.; Luxi, H.; Jian, J.; Yu, X. A Fast Iterative Pursuit Algorithm in Robust Face Recognition Based on Sparse Representation. *Math. Probl. Eng.* **2014**, *2014*.

27. Zhang, W.; Shan, S.; Chen, X.; Gao, W. Local Gabor binary patterns based on Kullback–Leibler divergence for partially occluded face recognition. *IEEE Signal Process. Lett.* **2007**, *14*, 875–878.

28. Reynolds, D.; Quatieri, T.; Dunn, R. Speaker verification using adapted Gaussian mixture models. *Digit. Signal Process.* **2000**, *10*, 19–41. [

29. Povey, D.; Chu, S.M.; Varadarajan, B. Universal background model based speech recognition. Proceedings of the IEEE International Conference on Acoustics, Speech and Signal Processing, Las Vegas, NV, USA, 30 March–4 April 2008; pp. 4561–4564.

30. Reynolds, D.A. An overview of automatic speaker recognition technology. Proceedings of the 2002 IEEE International Conference on Acoustics, Speech, and Signal Processing (ICASSP), Orlando, FL, USA, 13–17 May 2002; Volume 4, pp. 4072–4075.

31. Reynolds, D. Gaussian mixture models. In *Encyclopedia of Biometric Recognition*; Springer: New York, NY, USA, 2008; pp. 12–17.

32. Xiong, Z.; Zheng, T.; Song, Z.; Soong, F.; Wu, W. A tree-based kernel selection approach to efficient Gaussian mixture model–universal background model based speaker identification.*Speech Commun.* **2006**, *48*, 1273–1282.

33. Wang, B.; Li, W.; Yang, W.; Liao, Q. Illumination normalization based on weber's law with application to face recognition. *IEEE Signal Process. Lett.* **2011**, *18*, 462–465.

34. Lowe, D.G. Distinctive image features from scale-invariant keypoints. *Int. J. Comput. Vis.***2004**, *60*, 91–110.

35. Bosch, A.; Zisserman, A.; Muñoz, X. Scene classification via pLSA. *Lect. Notes Comput. Sci.***2006**, *3954*, 517–530.

36. Simonyan, K.; Parkhi, O.M.; Vedaldi, A.; Zisserman, A. Fisher vector faces in the wild. Proceedings of the British Machine Vision Conference, Bristol, UK, 9–13 September 2013; Volume 1, p. 7.

37. Shinoda, K.; Inoue, N. Reusing Speech Techniques for Video Semantic Indexing [Applications Corner]. *IEEE Signal Process. Mag.* **2013**, *30*, 118–122.

38. Ke, Y.; Sukthankar, R. PCA-SIFT: A more distinctive representation for local image descriptors. Proceedings of the 2004 IEEE Computer Society Conference on Computer Vision and Pattern Recognition, Washington, DC, USA, 27 June–2 July 2004; Volume 2, pp. 506–513.

39. Herbei, R.; Wegkamp, M.H. Classification with reject option. *Can. J. Stat.* **2006**, *34*, 709–721.

40. Samaria, F.S.; Harter, A.C. Parameterisation of a stochastic model for human face identification. Proceedings of the Second IEEE Workshop on Applications of Computer Vision, Sarasota, FL, USA, 5–7 December 1994; pp. 138–142.

41. Georghiades, A.; Belhumeur, P.; Kriegman, D. From Few to Many: Illumination Cone Models for Face Recognition under Variable Lighting and Pose. *IEEE Trans. Pattern Anal. Mach. Intell.* **2001**, *23*, 643–660.

42. Martinez, A.M. *The AR Face Database*; CVC Technical Report; The Ohio State University: Colombus, OH, USA, 1998; Volume 24.

43. Zhang, D.L.; Qiao, J.; Li, J.B.; Qiao, L.Y.; Chu, S.C.; Roddick, J.F. Optimizing Matrix Mapping with Data Dependent Kernel for Image Classification. *J. Inf. Hiding Multimed. Signal Process.* **2014**, *5*, 72–79.

44. Xia, S.; Zheng, G.; Ma, Y.; Ma, X. A. Novel Super-resolution Approach Based on Supervised Canonical Correlation Analysis. Proceedings of the 2014 International Conference on Computer, Communications and Information Technology (CCIT 2014), Beijing, China, 16–17 January 2014.

45. Xu, W.; Lee, E.J. Face Recognition Using Wavelets Transform and 2D PCA by SVM Classifier. *Int. J. Multimed. Ubiquitous Eng.* **2014**, *9*, 281–290.

46. Morelli Andrés, A.; Padovani, S.; Tepper, M.; Jacobo-Berlles, J. Face recognition on partially occluded images using compressed sensing. *Pattern Recognit. Lett.* **2014**, *36*, 235–242.

47. Yang, J.; Qian, J.; Luo, L.; Zhang, F.; Gao, Y. Nuclear Norm based Matrix Regression with Applications to Face Recognition with Occlusion and Illumination Changes. **2014**. arXiv preprint arXiv:1405.1207.

48. Jiang, Y.; Wu, Y.; Li, W.; Wang, L.; Liao, Q. Log-domain polynomial filters for illumination-robust face recognition. Proceedings of the 2014 IEEE International Conference on Acoustics, Speech and Signal Processing (ICASSP), Firenze, Italy, 4–9 May 2014; pp. 504–508.

49. Jia, Q.; Fang, C.; Wen, D.; Ding, X. Generating face images under multiple illuminations based on a single front-lighted sample without 3D models. Proceedings of the 2013 International Conference on Biometrics (ICB), Madrid, Spain, 4–7 June 2014; pp. 1–6.

50. Nabney, I.T. *NETLAB: Algorithms for Pattern Recognition*; Springer: London, UK, 2004.

51. Vedaldi, A.; Fulkerson, B. VLFeat: An open and portable library of computer vision algorithms. Proceedings of the International Conference on Multimedia, Firenze, Italy, 25–29 October 2010; pp. 1469–1472.

52. Miao, Z.; Jiang, X. Interest point detection using rank order LoG filter. *Pattern Recognit.* **2013**,*46*, 2890–2901.

53. Ross, A.; Rattani, A.; Tistarelli, M. Exploiting the "doddington zoo" effect in biometric fusion. Proceedings of the IEEE 3rd International Conference on Biometrics: Theory, Applications, and Systems 2009, Washington, DC, USA, 28–30 September 2009; pp. 1–7.

Chapter 10

Familiar and Unfamiliar Face Recognition in a Crowd

Hiroshi Ito[1], Aki Sakurai[2]

[1]Department of Psychology, Aichi University, Toyohashi, Japan

[2]Chubu Telecommunications Co., Nagoya, Japan

ABSTRACT

The purpose of the present study was to investigate whether the visual search process for familiar faces differs from that for unfamiliar faces. We used a single-target visual search task and recorded eye movements of participants during the task. We employed three different types of face familiarity: personally familiar faces (friends and teachers), famous faces, and unfamiliar faces. Participants had to search through arrays of faces for a target face. In each trial, a target face and three distractor faces were horizontally aligned and presented at the same time. In the personally familiar or famous condition, the target was a personally familiar face or a famous face, respectively, and the distractor faces were unfamiliar. In the unfamiliar condition, the target was an unfamiliar face and the distractor faces were personally familiar. The reaction times to identify the target demonstrated that the visual search for unfamiliar faces is slower than that for personally familiar faces and famous faces, but there was no significant difference between reaction times to recognize the personally familiar and the famous face targets. Additionally, the eye movement results in the unfamiliar face condition showed that an exhaustive search of the entire array occurred more frequently than a self-terminating search, but this was not true for both the personally familiar faces and famous faces conditions. These results suggest that the visual search process for familiar faces (personally familiar and famous) in a crowd differs from that for unfamiliar faces.

KEYWORDS

Personally Familiar Face, Famous Face, Unfamiliar Face, Face Perception, Visual Search

1. INTRODUCTION

People recognize familiar faces in a crowd. There has been a continuing interest in how people recognize familiar and unfamiliar faces since the 1970s (Bruce, 1979; Ellis, Shepherd, & Davies, 1979) to the present (Persike, Meinhardt-Injac, & Meinhardt, 2013; Walton & Hill, 2012). Many previous studies suggest that familiar and unfamiliar faces are processed in different ways (Balas, Cox, & Conwell, 2007; Barton, Radcliffe, Cherkasova, Edelman, & Intriligator, 2006; Clutterbuck & Johnston, 2002; Gobbini, Leibenluft, Santiago, & Haxby, 2004; Persike et al., 2013; Young, Hay, McWeeny, Flude, & Ellis, 1985; see Johnston & Edmonds, 2009; Natu & O'Toole, 2011, for review). Both behavioral and neuroscience studies have provided evidence to support the differences in the processing of familiar and unfamiliar faces.

Johnston and Edmonds (2009) pointed out that the major differences between familiar and unfamiliar face processing have been observed in behavioral experiments. Participants can recognize familiar faces in poor viewing conditions, such as in low-quality images and at variable viewpoints (Balas et al., 2007; Burton, Bruce, & Hancock, 1999; Johnston & Edmonds, 2009; Natu & O'Toole, 2011). Increased familiarity of a face results in increased accuracy of identification and facilitation of the recognition process under many kinds of incomplete recognition conditions. On the other hand, recognizing unfamiliar faces is easily disrupted in poor lighting conditions (Hill & Bruce, 1996; Johnston, Hill, & Carman, 1992), by changes in viewpoint (Balas et al., 2007; Bruce, 1982; Bruce et al., 1999; Hill, Schyns, & Akamatsu, 1997; O'Toole, Edelman, & Bulthoff, 1998), and when image or video quality is poor (Balase et al., 2007; O'Toole et al., 2010). It is also known that the internal features of a face (eyes, nose, and mouth area) are more important for the recognition of familiar faces than the external features (face shape and hair), although the recognition of unfamiliar faces is similarly efficient from internal or external features (Johnston & Edmonds, 2009; O'Donnell & Bruce, 2001; Young et al., 1985). For example, in an earlier study,Young et al. (1985) reported that familiar faces were matched faster when paired with internal rather than with external features, although unfamiliar face matches presented no such difference. Subsequent studies showed that unfamiliar faces were recognized more slowly and less accurately than familiar faces if the viewpoint was incongruent between the study phase and the test phase (Balas et al., 2007; Hill et al., 1997). The many advantages of familiar faces are probably derived from the differences in the neural mechanisms involved in the recognition of these faces.

Many neuropsychological studies have described patients who present various face recognition impairments, for example, prosopagnosia or other dissociations between performance of tasks that require the processing of familiar and unfamiliar faces (Malone, Morris, Kay, & Levin, 1982; Tranel, Damasion, & Damasio, 1988; Warrington & James, 1967; Young, Newcombe, DeHaan, Small, & Hay, 1993). For example, Young et al. (1993) described prosopagnosia patients who could identify familiar faces but were poor at identifying unfamiliar faces. These findings suggest that there are different neural mechanisms between recognition of familiar and unfamiliar faces.

Functional neuroimaging studies have detected some brain regions involved in unfamiliar face perception that respond more strongly to faces than to other objects (Natu & O'Toole, 2011). This network of brain regions consists of the fusiform gyrus (Hoffman & Haxby, 2000; Kanwisher, McDermott, & Chun, 1997), the inferior occipital gyrus (Halgren et al., 1999), and the superior temporal sulcus (STS) (Allison, Puce, & McCarthy, 2000; Haxby, Hoffman, & Gobbini, 2000). These core face-selective brain areas include the representation of the invariant aspects of faces for face recognition in the fusiform and inferior occipital gyri, and the variant aspects of faces, such as expression, gaze, and facial speech, in the posterior STS (Haxby et al., 2000; Natu & O'Toole, 2011). On the other hand, familiar faces activate not only the core face-selective regions, but also other brain regions that represent semantic, episodic, and emotional information about an individual (Ewbank & Andrews, 2008; Gobbini et al., 2004; Gobbini & Haxby, 2007; Gorno Tempini & Price, 2001; Rossion, Schiltz, & Crommelinck, 2003; Sugiura et al., 2001). These richly interconnected representations could affect the ease of recognition of familiar faces.

Although it is widely acknowledged that there are differences between the processing of familiar and unfamiliar faces, little is known about the visual search process for a familiar face in a crowd of unfamiliar faces and vice versa. The aim of the present study was to examine whether the visual search process for familiar faces differs from that for unfamiliar faces. In order to assess this issue, we used a single-target visual search task (Becker, Anderson, Mortensen,

Neufeld, & Neel, 2011;Treisman & Gelade, 1980; Thornton & Gilden, 2007; Wolfe, 1992) and recorded eye movements of participants during the task by using an eye tracker. This task requires participants to search through arrays of objects for a target object and, therefore, can disclose differences in the efficiency with which particular target classes can be found (Becker et al., 2011). In the present study, we used only target-present trials, that is, targets were presented in every array. If a target is equally likely to be present in each possible location within an array, participants would not need to search the arrays exhaustively (an exhaustive search), but would terminate the search when the target is found (a self-terminating search).

We were also interested in investigating the effect of different types of face familiarity: personally familiar faces (e.g., close relatives and friends) known through real-world experiences, and famous faces (e.g., celebrities and politicians) whom the participants knew through the media. Previous studies suggested that personal familiarity of faces differs from famous face familiarity in terms of the amount of exposure, knowledge about personal traits, and emotional and semantic associations (Gobbini et al., 2004; Natu & O'Toole, 2011). Thus, personal familiarity is associated with not only the visual information of a face, but also with emotional and social information, such as the person's mental state, personality, and the emotions that people experience upon seeing the face (Gobbini & Haxby, 2007).

2. METHOD

2.1. Participants

Twelve undergraduate students (3 males and 9 females; age range: 21 to 23 years) participated in the present experiment. All participants had normal or corrected-to-normal vision and were naive with respect to the hypotheses under investigation. They gave written informed consent according to the recommendations of the Declaration of Helsinki to participate in the experiment. This study was approved by the local ethics committee.

2.2. Design

The experiment employed a one-factor (level of face familiarity: personally familiar face, famous face, or unfamiliar face) within-subject design.

2.3. Stimuli

We used gray-scale photographs of faces with a high-resolution frontal view so that broad color cues could not facilitate recognition of targets. Figure 1 illustrates examples of the images used in this experiment. We rejected images with excess jewelry, make-up, open mouths, or glasses. The images were 224 × 192 pixels in size. Each face was isolated from the original photographs and superimposed on a gray background using image editing software (Photoshop CS6; Adobe, California, USA). We employed three different familiarities of faces: personally familiar faces (friends and teachers in the same department of psychology), famous faces (actors, singers, public leaders, athletes), and unfamiliar faces (anonymous individuals).

For personally familiar faces, participants in this experiment were shown the entire set of 70 individuals (28 males and 42 females) who are undergraduate students or faculty in the Department of Psychology at Aichi University, and were asked to select 32 individuals familiar to them a month before the experiment. Eight of the selected individuals were randomly assigned to be target stimuli in the personally familiar condition, and the remaining 24 of them were assigned to be distractor stimuli in the unfamiliar condition. The participants rated these faces for familiarity on a 10-point scale, with 0 being not familiar at all and 10 being extremely familiar to them. The average familiarity rating for all personally familiar faces was 8.1 (SD 1.2). For famous faces, we selected eight images from the internet (4 males and 4 females). The eight famous faces were only assigned to be target stimuli in the famous condition and were not assigned to be distractor stimuli in any other conditions. All the famous faces were well known to the participants. The participants rated these faces for familiarity on a 10- point scale in the same manner as the personally familiar faces. The average familiarity rating for all famous faces was 8.7 (SD 0.8). For unfamiliar faces, we used 56 individuals who were selected from the pool of photographs of students and faculty in other university (28 males and 28 females). Eight of the selected individuals were randomly assigned to be target stimuli in the unfamiliar condition, and the remaining 48 of them were assigned to be distractor stimuli. The half of the distractor stimuli (24 individuals) was used in the personally familiar condition, and the remaining half of them (24 individuals) was used in the famous condition. These unfamiliar faces were selected to match personally familiar and famous faces on age, race, and gender. In total we presented participants with 24 target stimuli (8 targets × 3 conditions) and 72 distractor stimuli (24 distractors × 3 conditions). None of the target and distractor stimuli were used more than once in each participant.

In the visual search task in the present study, a target face (personally familiar, famous, or unfamiliar) and three distractor faces were horizontally aligned and presented at the same time. In the personally familiar or famous condition, the target face was a personally familiar face or a famous face, respectively, and the distractor faces were unfamiliar faces. Conversely, in the unfamiliar condition, the target face was an unfamiliar face and the distractor faces were personally familiar faces. None of the target stimuli were used more than once in each condition.

(a) (b) (c)

Figure 1. Examples of stimuli used in the present study. (a) Personally familiar faces; (b) Famous faces; (c) Unfamiliar faces.

2.4. Apparatus

The stimuli were displayed on a 17-inch CRT color monitor (EDT17AD, EPSON, Suwa, Japan) with a screen resolution of 1280 × 1024 pixels and a refresh rate of 75 Hz. This monitor was located at a distance of approximately 50 cm away from the participant. The mean luminance of the screen was 75 cd/m^2. Eye movements were recorded using an eye tracker (EMR-AT VOXER, NAC Image Technology Inc., Tokyo, Japan) with a sampling rate of 60 Hz. We used EMR-dFactory version 1.1, eye movement data analysis software (NAC Image Technology Inc., Tokyo Japan). The eye tracker was calibrated to each participant's dominant eye, but viewing was binocular. Calibration was done using a nine-point calibration accuracy test. A chin rest maintained the participants' viewing position and distance.

2.5. Procedure

The participants were tested individually in a dimly lit room. The room was darkened so that the ambient illumination approximately matched the illumination on the screen. They were comfortably seated and performed the visual search task. In this task, they were asked to indicate a target face in a crowd of three distractor faces. They were encouraged to respond as quickly and accurately as possible. We recorded their eye movements during the task.

Each trial was begun by presenting a black fixation point displayed at the center of the monitor on a white background for one second, followed by a blank interval of 0.5 seconds. Then, a target face and three distractor faces were presented until the participant responded. These four faces were horizontally aligned and presented at the same time. The centers of the rightmost and leftmost face images were about 12 cm (442 pixels) away from the center of the screen, and the centers of the two face images at the middle of the face array were about 4 cm (147 pixels) away from the center of the screen. The task of the participants was to look for an instructed target face (personally familiar face, famous face, or unfamiliar face). They were asked to press the "1" key with the left middle finger when the target face was presented at the leftmost position, to press the "2" key with the left index finger when it was presented at the second position from the left, to press the "3" key with the right index finger when it was presented at the second position from the right, and to press the "4" key with the right middle finger when it was presented at the rightmost position. After responding, participants were given an intertrial interval of five seconds. Participants were not given feedback on whether their response was correct or not.

Prior to participating in the experimental trials, each participant completed a block of 6 practice trials using the experimenters' faces. The experimental session consisted of three blocks (the personally familiar face condition, the famous face condition, and the unfamiliar face condition). The order of the three experimental blocks was counterbalanced across participants, and each block consisted of eight trials. In total we presented each participant with 24 trials consisting in three blocks of eight trials. The target faces were presented once in one of the four positions in each block. The number of presented target faces was balanced across the four positions. The order of the eight trials in each condition was randomized across participants. After each experimental block there was a two-minute rest period. The experimental session lasted for approximately ten minutes.

After the visual search task, all participants took part in a familiarity verification exercise. They were shown printed copies of all the unfamiliar faces and asked to provide any copies of the unfamiliar faces if they had seen these individuals before. As a result, none of the participants had ever seen these individuals before.

3. RESULTS

3.1. Accuracy

The average performance of correct responses for all participants across all three conditions exceeded 99% (personally familiar: 99%; famous: 99%; unfamiliar: 100%). In order to assess whether the familiarity of faces affects the visual search accuracy of a target face in a crowd of distractors, the proportion of correct responses was entered into a one-way repeated-measures ANOVA with the level of face familiarity (personally familiar, famous, or unfamiliar) as the within-subject factor. We found no significant main effect of the level of face familiarity, $F < 1$ between the proportions of correct responses. This result indicates that participants accurately found target faces in the crowd of distractor faces in all three conditions.

3.2. Reaction Time

Mean reaction times for correct responses to target in each experimental condition are given inFigure 2. In order to reduce outlier effects (more than three standard deviations above the mean value for the condition), responses above three seconds (1.1% of all correct responses) were excluded from the analyses. A one-way repeated-measures ANOVA of the reaction times for correct responses was carried out to determine the effect of face familiarity. We found a significant main effect of the level of face familiarity on reaction time, $F(2, 22) = 4.48$, $p < .05$. Post-hoc comparisons using Ryan's procedure ($\alpha = .05$) revealed that reaction times in the unfamiliar face condition (1686 ms) were longer than those in the personally familiar face (1394 ms) or famous face (1418 ms) conditions ($p < .05$), although there was no significant difference of reaction times between the personally familiar face and the famous face conditions. These results suggest that the visual search for unfamiliar faces is slower than that for personally familiar faces or famous faces.

3.3. Eye Movements

We recorded the participants' eye movement during the visual search task. In order to examine the difference in the visual search strategies used to recognize familiar faces versus unfamiliar faces, we focused on the proportions of those using a self-terminating search or an exhaustive search for the visual target. In this experiment, the self-terminating search was defined as the search strategy in which participants stop searching without seeing all four faces presented in a trial, and the exhaustive search was defined as the search strategy in which the participants see all the faces. Based on eye fixations that defined the intervals between the end of one saccade and the beginning of the next, we examined whether these fixations landed on the areas of interest that spanned from the top of the head to the bottom of the chin of each of the four faces presented in each trial.

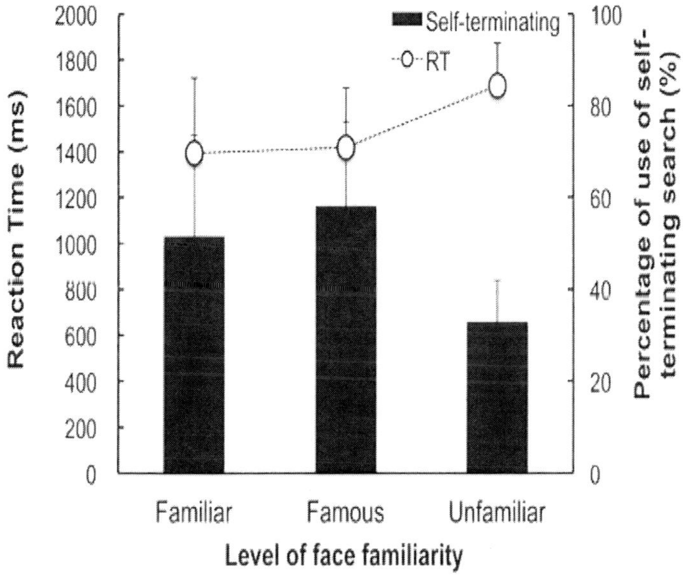

Figure 2. Mean reaction times (ms) for correct target searching and mean percentage of use of a self-terminating search in the personally familiar face, the famous face, and the unfamiliar face conditions. The open circles indicate the reaction times, and the solid bars indicate the percentages of use of the selfterminating search. Error bars represent standard error of the mean.

Figure 2 also shows the average proportion of participants who used the self-terminating search in each experimental condition. A one-way repeated-measures ANOVA of the percentage of participants using the self-terminating search was carried out to determine the effect of face familiarity (personally familiar, famous, or unfamiliar). We found a significant main effect of the level of face familiarity, $F(2, 22) = 5.48$, $p < .05$. Post-hoc comparisons using Ryan's procedure ($\alpha = .05$) revealed that the percentages of self-terminating search used in the unfamiliar face condition (33%) were lower than those used in the personally familiar face (52%) and famous face (58%) conditions ($p < .05$), although there was no significant difference in percentage of self-terminating search used between the personally familiar face and the famous face conditions. These results suggest that use of the self-terminating search in the unfamiliar face condition occurred less frequently than in the personally familiar face and famous face conditions, and that the exhaustive search is more common than the selfterminating search in the unfamiliar face condition.

4. DISCUSSION

The purpose of the present study was to investigate whether the visual search process for familiar faces differs from that for unfamiliar faces. In order to address this issue, the experiment was conducted by using a singletarget visual search task for each face condition while recording eye movements during the task. We employed three different levels of face familiarity: personally familiar faces, famous faces, or unfamiliar faces. The mean reaction times demonstrate that the visual search for unfamiliar faces is slower than that for personally familiar faces and famous faces, but there is no significant difference in the time taken to recognize personally familiar and famous faces. Eye movements during the visual search task showed that the exhaustive search occurred more frequently than the self-terminating search when participants searched through arrays of personally familiar faces for a unfamiliar target face, but both personally familiar face and famous face target searches did not show more frequent use of the exhaustive search.

Our findings illuminate the mechanisms underlying the visual search for familiar and unfamiliar faces in terms of search facilitation and strategy. The reaction time results suggest that both the real world experiences with faces (i.e., personally familiar faces) and experience through the media (i.e., famous faces) facilitate the processing of facial identification in a crowd. These results are consistent with previous studies that have shown advantages in recognizing familiar faces over unfamiliar faces (e.g., Balas et al., 2007; Clutterbuck & Johnston, 2002; Young et al., 1985). Visual components of rich internal features (e.g., eyes and mouth) and semantic longterm memory components (e.g., biographical information, name, and personal traits) of familiar faces allow for faster recognition than unfamiliar faces. Although it is not clear whether the face processing advantages observed in the personally familiar and famous faces arise from the same basic mechanism or from separate ones, a variety of mechanisms could cause fast processing of familiar faces. One possibility is that the internal features of often-viewed faces induce faster focusing on particular features of a target face (e.g., the shape of the eyes) (Balas et al., 2007; Young et al., 1985) or on configural information about the spatial relationships between the internal features of the face (Johnston & Edmonds, 2009).

Another possibility is that a strong facial memory provides a low threshold to be able to make a decision about a familiar face target. The prior observation of those faces would function as a visual priming effect. However, the present study cannot distinguish between these possibilities.

Interestingly, the exhaustive search for a target face occurred more frequently than the self-terminating search in the unfamiliar face condition, but this was not true in both the personally familiar face and the famous face conditions. These results imply that people do an exhaustive search of the entire array before they indicate the position of an unfamiliar face. Previous studies have suggested that there is a longer reaction time involved in the recognition of unfamiliar faces (e.g., Balas et al., 2007; Clutterbuck & Johnston, 2002; Young et al., 1985) and a longer duration of scanning of unfamiliar faces before the formation of a perceptual decision (Barton et al., 2006). However, in a visual search for unfamiliar faces, our results suggest that a longer reaction time for unfamiliar face recognition in a crowd results not only from the longer duration of scanning a target unfamiliar face, but also from the exhaustive search of the entire array. A plausible interpretation of this scanning pattern is that participants compare an unfamiliar face (target) with familiar faces (distractors) if the unfamiliar face itself is not enough to make a correct decision. Thus, the participants defer a decision on an unfamiliar target face until the scanning for familiar faces is over. In terms of limitations, the relatively small sample size (n = 12) in the current study means that our results should be interpreted with caution.

In conclusion, our study demonstrates that the visual search process for familiar faces (i.e., personally familiar and famous) differs from that for unfamiliar faces. Our results indicate that an exhaustive search of entire arrays occurred more frequently than a self-terminating search when participants searched through arrays of personally familiar faces for an unfamiliar face, but this was not true for both personally familiar faces and famous faces searches.

AUTHORSHIP

Ito and Sakurai developed the study concept. Both authors contributed to the study design. Testing and data collection were performed by Ito and Sakurai, and Ito and Sakurai performed the data analysis and interpretation. Ito wrote the paper. All authors approved the final version of the paper for submission.

REFERENCES

1. Allison, T., Puce, A., & McCarthy, G. (2000). Social Perception from Visual Cues: Role of the STS Region. Trends in Cognitive Sciences, 4, 267-276.

2. Balas, B., Cox, D., & Conwell, E. (2007). The Effect of Real-World Personal Familiarity on the Speed of Face Information Processing. PLoS ONE, 2, Article ID: e1223.

3. Barton, J. J. S., Radcliffe, N., Cherkasova, M. V., Edelman, J., & Intriligator, J. M. (2006). Information Processing during Face Recognition: The Effects of Familiarity, Inversion, and Morphing on Scanning Fixations. Perception, 35, 1089-1105.

4. Becker, D. V., Anderson, U. S., Mortensen, C. R., Neufeld, S. L., & Neel, R. (2011). The Face in the Crowd Effect Unconfounded: Happy Faces, Not Angry Faces, Are More Efficiently Detected in Singleand Multiple-Target Visual Search Tasks. Journal of Experimental Psychology: General, 140, 637-659.

5. Bruce, V. (1979). Searching for Politicians: An Information Processing Approach to Face Recognition. Quarterly Journal of Experimental Psychology, 31, 373-395.

6. Bruce, V. (1982). Changing Faces: Visual and Non-Visual Coding Processes in Face Recognition. British Journal of Psychology, 73, 105-116.

7. Bruce, V., Henderson, Z., Greenwood, K., Hancock, P. J. B., Burton, A. M., & Miller, P. (1999). Verification of Face Identities from Images Captured on Video. Journal of Experimental Psychology: Applied, 5, 339-360.

8. Burton, A. M., Bruce, V., & Hancock, P. J. B. (1999). From Pixels to People: A Model of Familiar Face Recognition. Cognitive Science, 23, 1-31.

9. Clutterbuck, R., & Johnston, R. A. (2002). Exploring Levels of Face Familiarity by Using an Indirect Face-Matching Measure. Perception, 31, 985-994.

10. Ellis, H. D., Shepherd, J. W., & Davies, G. M. (1979). Identification of Familiar and Unfamiliar Faces from Internal and External Features; Some Implications for Theories of Face Recognition. Perception, 8, 431-439.

11. Ewbank, M. P., & Andrews, T. J. (2008). Differential Sensitivity for Viewpoint between Familiar and Unfamiliar Faces in Human Visual Cortex. NeuroImage, 40, 1857-1870.

12. Gobbini, M. I., & Haxby, J. V. (2007). Neural Systems for Recognition of Familiar Faces. Neuropsychologia, 45, 32-41.

13. Gobbini, M. I., Leibenluft, E., Santiago, N., & Haxby, J. V. (2004). Social and Emotional Attachment in the Neural Representation of Faces. NeuroImage, 22, 1628-1635.

14. Gorno-Tempini, M. L., & Price, C. J. (2001). Identification of Famous Faces and Buildings: A Functional Neuroimaging Study of Semantically Unique Items. Brain, 124, 2087-2097.

15. Halgren, E., Dale, A. M., Sereno, M. I., Tootell, R. B. H., Marinkovic, K., & Rosen, B. R. (1999). Location of Human Face-Selective Cortex with Respect to Retinotopic Areas. Human Brain Mapping, 7, 29-37.

16. Haxby, J. V., Hoffman, E. A., & Gobbini, M. I. (2000). The Distributed Human Neural System for Face Perception. Trends in Cognitive Sciences, 4, 223-233.

17. Hill, H., & Bruce, V. (1996). Effects of Lighting on the Perception of Facial Surfaces. Journal of Experimental Psychology: Human Perception and Performance, 22, 986-1004.

18. Hill, H., Schyns, P. G., & Akamatsu, S. (1997). Information and Viewpoint Dependence in Face Recognition. Cognition, 62, 201-222.

19. Hoffman, E. A., & Haxby, J. V. (2000). Distinct Representations of Eye Gaze and Identity in the Distributed Human Neural System for Face Perception. Nature Neuroscience, 3, 80-84.

20. Johnston, A., Hill, H., & Carman, N. (1992). Recognizing Faces: Effects of Lighting Direction, Inversion, and Brightness Reversal. Perception, 21, 365-375.

21. Johnston, R., & Edmonds, A. (2009). Familiar and Unfamiliar Face Recognition: A Review. Memory, 17, 577-596.

22. Kanwisher, N., McDermott, J., & Chun, M. M. (1997). The Fusiform Face Area: A Module in Human Extrastriate Cortex Specialized for Face Perception. Journal of Neuroscience, 17, 4302-4311.

23. Malone, D. R., Morris, H. H., Kay, M. C., & Levin, H. S. (1982). Prosopagnosia: A Double Dissociation between the Recognition of Familiar and Unfamiliar Faces. Journal of Neurology, Neurosurgery & Psychiatry, 45, 820-822.

24. Natu, V., & O'Toole, A. J. (2011). The Neural Processing of Familiar and Unfamiliar Faces: A Review and Synopsis. British Journal of Psychology, 102, 726-747.

25. O'Donnell, C., & Bruce, V. (2001). Familiarisation with Faces Selectively Enhances Sensitivity to Changes Made to the Eyes. Perception, 30, 755-754.

26. O'Toole, A. J., Edelman, S., & Bülthoff, H. H. (1998). Stimulus-Specific Effects in Face Recognition over Changes in Viewpoint. Vision Research, 38, 2351-2363.

27. O'Toole, A. J., Weimer, S., Dunlop, J., Barwick, R., Ayyad, J., & Phillips, P. J. (2010). Recognizing People from Dynamic Video: Dissecting Identity with a Fusion Approach. Vision Research, 10, 643.

28. Persike, M., Meinhardt-Injac, B., & Meinhardt, G. (2013). The Preview Benefit for Familiar and Unfamiliar Faces. Vision Research, 87, 1-9.

29. Rossion, B., Schiltz, C., & Crommelinck, M. (2003). The Functionally Defined Right Occipital and Fusiform "Face Areas" Discriminate from Visually Familiar Faces. NeuroImage, 19, 877-883.

30. Sugiura, M., Kawashima, R., Nakamura, K., Sato, N., Nakamura, A., Kato, T., & Fukuda, H. (2001). Activation Reduction in Anterior Temporal Cortices during Repeated Recognition of Faces of Personal Acquaintances. NeuroImage, 13, 877-890.

31. Thornton, T. L., & Gilden, D. L. (2007). Parallel and Serial Processes in Visual Search. Psychological Review, 114, 71-103.

32. Tranel, D., Damasio, A. R., & Damasio, H. (1988). Intact Recognition of Facial Expression, Sex, and Age in Patients with Impaired Recognition of Facial Identity. Neurology, 28, 690-696.

33. Treisman, A. M., & Gelade, G. (1980). A Feature-Integration Theory of Attention. Cognitive Psychology, 12, 97-136.

34. Walton, B. R. P., & Hills, P. J. (2012). Face Distortion Aftereffects in Personally Familiar, Famous, and Unfamiliar Faces. Frontiers in Psychology, 3, Article ID: 258.

35. Warrington, E. K., & James, M. (1967). An Experimental Investigation of Facial Recognition in Patients with Unilateral Cerebral Lesions. Cortex, 3, 317-326.

36. Wolfe, J. M. (1992). The Parallel Guidance of Visual Attention. Current Directions in Psychological Science, 1, 124-128.

37. Young, A. W., Hay, D. C., McWeeny, K. H., Flude, B. M., & Ellis, A. W. (1985). Matching Familiar and Unfamiliar Faces on Internal and External Features. Perception, 14, 737-746.

38. Young, A. W., Newcombe, F., DeHaan, E. H. F., Small, M., & Hay, D. C. (1993). Face Perception after Brain Injury. Selective Impairments Affecting Identity and Expression. Brain, 116, 941-959.

Chapter 11

An automated chimpanzee identification system using face detection and recognition

Alexander Loos1* and Andreas Ernst2

1Audio-Visual Systems, Fraunhofer IDMT, 98693 Ilmenau, Germany.

2Electronic Imaging, Fraunhofer IIS, 91058 Erlangen, Germany.

ABSTRACT

Due to the ongoing biodiversity crisis, many species including great apes like chimpanzees are on the brink of extinction. Consequently, there is an urgent need to protect the remaining populations of threatened species. To overcome the catastrophic decline of biodiversity, biologists and gamekeepers recently started to use remote cameras and recording devices for wildlife monitoring in order to estimate the size of remaining populations. However, the manual analysis of the resulting image and video material is extremely tedious, time consuming, and cost intensive. To overcome the burden of time-consuming routine work, we have recently started to develop computer vision algorithms for automated chimpanzee detection and identification of individuals. Based on the assumption that humans and great apes share similar properties of the face, we proposed to adapt and extend face detection and recognition algorithms, originally developed to recognize humans, for chimpanzee identification. In this paper we do not only summarize our earlier work in the field, we also extend our previous approaches towards a more robust system which is less prone to difficult lighting situations, various poses, and expressions as well as partial occlusion by branches, leafs, or other individuals. To overcome the limitations of our previous work, we combine holistic global features and locally extracted descriptors using a decision fusion scheme. We present an automated framework for photo identification of chimpanzees including face detection, face alignment, and face recognition. We thoroughly evaluate our proposed algorithms on two datasets of captive and free-living chimpanzee individuals which were annotated by experts. In three experiments we show that the presented framework outperforms previous approaches in the field of great ape identification and achieves promising results. Therefore, our system can be used by biologists, researchers, and gamekeepers to estimate population sizes faster and more precisely than the current frameworks. Thus, the proposed framework for chimpanzee identification has the potential to open up new venues in efficient wildlife monitoring and can help researches to develop innovative protection schemes in the future.

KEYWORDS

Wildlife monitoring, African great apes, Face and facial feature detection, Individual identification

1 INTRODUCTION

According to the International Union for Conservation of Nature (IUCN), about 22% of the mammal species worldwide are threatened or extinct [1]. The current biodiversity crisis is observed all over the world. Primates are hit by the crisis and belong to a species that is severely endangered. Walsh et al. [2] reported a decrease of ape populations in western Equatorial Africa by more than a half between 1983 and 2000. A similar survey was done by Campbell et al. [3]. They observed a 90% decrease of chimpanzee sleeping nests in Côte d'Ivoire between 1990 and 2007.

Those agitating results demonstrate the urgent need to intensify close surveillance of this threatened species. Many protective areas have already been established. However, effectively protecting the animals requires a good knowledge of existing populations and changes of population sizes over time. Individual identification of animals is not only a prerequisite for measuring the success of implemented protection schemes but also for many other biological questions, e.g., wildlife epidemiology and social network analysis. However, it is a labor-intensive task to estimate population sizes in the wild. Therefore, noninvasive monitoring techniques that take advantage of automatic camera traps are currently under development, and the number of published studies that use camera traps or autonomous recording devices is highly increasing [4]. However, the collected data are still evaluated manually which is a time- and resource-consuming task. Consequently, there is a high demand for automated algorithms to analyze remotely gathered video recordings. Especially so-called capture-mark-recapture methods, commonly used in ecology, could benefit from an automated system for identification of great apes.

This paper shows that technology developed for human face detection and identification can provide substantial assistance in evaluating data gathered by camera traps. We summarize and extend our previous work from [5]- [9] on face detection and individual identification of African great apes for wildlife monitoring and present an automated framework to detect and subsequently identify free-living as well as captured chimpanzee individuals in uncontrolled environments.

Some aspects of this paper have been published in our previous work. We extended our approaches from [6] and [7] to improve the system's robustness against pose variations, difficult lighting conditions, and partial occlusions [8]. However, in this paper we present a complete system for chimpanzee photo identification including face detection, face alignment, and face recognition. We significantly improve previous approaches by fusing global and local descriptors in a decision-based manner.

While global descriptors represent the whole appearance of a chimpanzee's face, the local features around certain facial fiducial points are more robust against local changes as they only encode detailed traits of the corresponding point of interest. Furthermore, it is well known from psychophysics and neuroscience that both holistic and local information are crucial for perception and recognition of faces. Starting from the assumption that a combination of global and local descriptors should improve the performance and robustness of the system, we use a decision fusion scheme to combine their strengths. We show that global feature vectors obtained by Gabor features in combination with speeded-up robust features (SURF) [10] as local face representation achieve promising results in the new field of face recognition of great apes and clearly outperform the system presented in our previous work. For evaluation we use two realistic real-world datasets of chimpanzees, gathered in the zoo and in the field. In summary, this paper contains three main contributions:

Presentation of an automated framework for primate photo identification including face detection, face alignment and lighting normalization, as well as identification.

1 Extension and improvement of our previous work to achieve better performance and more robustness against pose variation, lighting conditions, facial expressions, noncooperative subjects, and even partial occlusion by branches or leafs.

2 Evaluation of the proposed system on two realistic real-world datasets of free-living and captured chimpanzee individuals gathered in uncontrolled environments.

3 The outcome of this paper builds the basis of an automated system for primate identification in photos and videos, which could open up new venues in efficient wildlife monitoring and biodiversity conservation management.

The remaining paper is organized as follows: In the subsequent section, we give a short recap of the existing work in the field of animal detection and identification and our own previous work. A detailed description of the proposed system, including face and facial feature detection, face alignment, and individual identification is presented in Section 3. We thoroughly evaluate our system on two datasets of free-living and captive chimpanzees in Section 4 using an open-set identification scheme. Finally, in Section 5, we conclude this paper and give further ideas of improvement.

2 RELATED WORK

The field of computer vision and pattern recognition has been an active research field for years. Even though automatic image and video processing techniques become more and more important for the detection and identification of animals, only few publications do exist dealing with that topic. In this section we give a brief overview of the existing technologies for the detection and identification of animals and briefly review face detection and recognition technologies developed for human identification.

2.1 Visual detection

Automatic face detection has been an important research area for many years now and has extensively been done for human faces. Rowley et al. [11] published good results with a neural network-based face detector more than 10 years ago. However, the system was not real-time capable at that time. Some years later, Viola and Jones [12] developed and published the probably best-known algorithm for real-time object detection. It uses AdaBoost [13] for feature selection and learning and benefits from the integral image to extract Haar-like features very fast. Numerous improvements and variants have been published in the literature afterwards [14]- [16].

Whereas plenty of work has already been done in the field of human face detection, only few publications can be found that deal with automatic detection, tracking, and analysis of animals in videos. Wawerla et al. [17] describe a system to monitor the behavior of grizzly bears at the arctic circle with camera traps. They use motion shapelet features and AdaBoost to detect bears in video footage. Burghardt and Calic [18] worked on the detection and tracking of animal faces based on the Viola-Jones detector and a low-level feature tracker. They trained the system on lion faces and showed that the results can be used to classify basic locomotive actions. Spampinato et al. [19,20] proposed a system for fish detection, tracking, and species classification in natural underwater environment. They first detect fishes using a combination of a Gaussian mixture model and moving average algorithms. The detected objects are then tracked using an adaptive mean shift algorithm. Finally, species classification is performed by combining texture and shape features to a powerful descriptor.

2.2 Visual identification

One of the most established and well-studied approaches for face recognition are appearance-based methods. Here the two-dimensional gray-level images with size $w \times h$ are represented as vectors of size $n = w \cdot h$. Thus, often simple pixel-based features are used as face descriptors. Since this high-dimensional feature space is too large to perform fast and robust face recognition in practice, dimensionality reduction techniques like principal component analysis (PCA) [21], linear discriminant analysis (LDA) [22], or locality preserving projections (LPP) [23] can be used to project the vectorized face images into a smaller dimensional subspace. These methods are often referred to as Eigenfaces, Fisherfaces, and Laplacianfaces, respectively. Recently, a random projection has also been successfully used for face recognition in combination with a sparse representation classification (SRC) scheme [24]. Random projection matrices can simply be generated by sampling zero-mean independent identically distributed Gaussian entries. This approach was extended by [25]. The authors suggest to use Gabor features instead of pixel-based features, which greatly improve the recognition accuracy while at the same time reduce the computational cost when dealing with occluded face images.

While biometric identification of humans has been an active research topic for decades, individual recognition of animals has only been addressed in the recent past. Ardovini et al. [26] for instance proposed a system for semiautomatic recognition of elephants from photos based on shape comparison of the nicks characterizing the elephant's ears. A similar approach was presented by Araabi et al. [27], who proposed a string matching method as part of a computer-assisted system for dolphin identification from images of their dorsal fin. Also Burghardt et al. [28,29] presented a fully automatic system for penguin identification. After a penguin has been detected, unique individual-specific spot patterns on the penguin's coat are used for identification. More recently a method called StripeCodes for zebra identification was published by Lahiri et al. [30]. The authors claim that their algorithm efficiently extracts simple image features used for the comparison of zebra images to determine whether the animal has been observed before or not.

To the best of our knowledge, the problem of nonhuman primate identification has not yet been addressed by other researchers so far.

2.3 Own work

The aforementioned approaches use characteristic coat patterns or other individually unique biometrics like the pattern of fur and skin as well as unique nicks in ears or dorsal fins to distinguish between individuals. Unfortunately, such an approach is often infeasible for the identification of great apes since unique coat markings are not existent or cannot be used because of the limited resolution of video recordings.

Based on the assumption that humans and our closest relatives share similar properties of the face, we suggested to use and adapt face recognition techniques, originally developed to recognize humans, for the identification of great apes within the SAISBECO project (http://www.saisbeco.com). In [5] we showed that state-of-the-art face recognition techniques are capable to also identify chimpanzees and gorillas. Based on these results, we significantly improved the performance of the proposed system by using Gabor features in combination with LPP for dimensionality reduction in [6]. The SRC scheme was used to assign identities to the facial images. Although the results of [6] are very promising, the accuracy of the system drops significantly if nonfrontal face images are used for testing. Another drawback is our assumption that faces and facial feature points were already detected properly for alignment and recognition. We overcame the latter issue by combining face and facial feature detection as well as face recognition and presented an automated identification system for chimpanzees in [7]. However, we only used simple pixel information in the recognition part of the proposed system. Thus although the achieved results were very promising for a first approach, the accuracy of the system was limited due to the lack of robustness against difficult lighting situations, pose, partial occlusion, and the various number of occurring expressions.

In this paper we show how to overcome this limitation by using more sophisticated face descriptors in combination with a powerful feature space transformation technique. By combining global and local features, the system's performance and robustness against above-mentioned situations can be further increased [8]. However, this technique has never been used within a complete identification framework for great apes including face detection, face alignment, and face recognition. Therefore, in this paper we propose, design, and evaluate an automated face detection and recognition system for chimpanzees in wildlife environments.

3 PROPOSED SYSTEM

Figure 1 gives an overview of the proposed system. It comprises three main components. In the first step chimpanzee faces in images are found and the eyes are located within the face regions. In the second component we apply several pre-processing steps like face alignment and lighting normalization to ensure comparability of the facial images across the entire database and improve the systems robustness against lighting changes. The third and last step recognizes the detected and normalized faces and assigns identities to them. The following subsections explain those three parts in more detail.

3.1 Face and facial feature detection

Detection of primate faces and localization of the facial feature points in images are necessary for the following individual identification. Our system uses detection models with multiple classification stages of increasing complexity as shown in Figure 2. Each stage can early reject a classification window as a nonface to decrease the computational complexity. All stages comprise a feature extractor and a classifier. The feature extractors use one out of three feature types. The number of features that are used in each stage is selected empirically. The first stages of the model use few simple features that can be calculated very fast and enable real-time processing. Subsequent stages apply more complex and distinctive features. They have only minor impact on the processing speed because only few classification windows reach those complex stages. Each classifier consists of look-up tables that have been built in an off-line training procedure using Real-AdaBoost [31]. Our system uses three types of features that are illumination invariant and robust against various lighting conditions. All features are solely based on gray value images and thus enable the system to process infrared images as well.

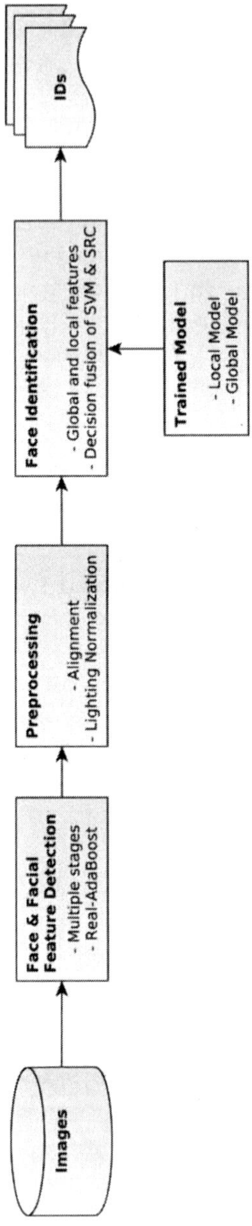

Figure 1. Overview of the proposed system. The figure depicts the overview of the proposed system for chimpanzee identification. After all possible faces in an image are detected, each face is aligned using a projective transform to make the faces comparable across the entire dataset. A histogram equalization is also applied in this step to improve the system's robustness against changing lighting conditions. In the final stage of our framework, the detected and aligned faces are identified using a combination of global and local features.

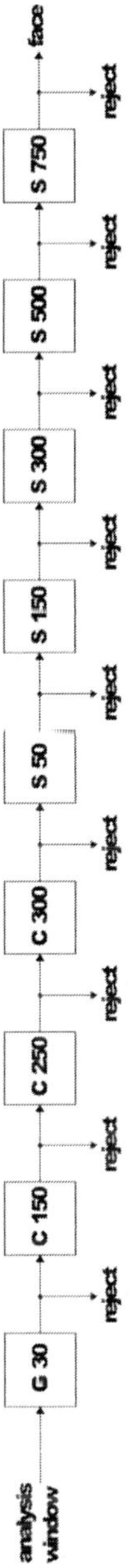

Figure 2. Setup of the face detection model. The face detection model comprises nine stages with different feature types and feature counts. The letters abbreviate 'G' for gradient, 'C' for census and 'S' for structure features. The first stages apply fast pixel-based features for the candidate search; the last stages use structure features for verification.

The first feature type describes the local gradient direction. Sobel kernels of size 3 × 3 extract the gradient s_x and s_y in *x*- and *y*-direction, similar to [32]. In homogeneous regions where s_x and s_y equals 0, the final feature is encoded as 0; otherwise, the feature encodes the result of atan2(s_y,s_x) quantized to the range 1...*q*. Experiments indicated that 35 is a good choice for *q* and results in a quantization interval of slightly more than 10°. We use*census* features [33] (also known as local binary patterns [34]) as a second feature type. These features describe the local brightness changes within a 3 × 3 neighborhood. The center pixel intensity is compared with its eight neighbors. The result is encoded as an 8-bit string that shows which neighboring pixels are less bright than the center pixel. The 3 × 3 local features are complemented by the third feature type that includes enlarged areas. Therefore, we encode structures by resized versions of census features that are calculated on image regions of 3*u* × 3*v* pixels. These *structure* features are a superset of the census features. Nevertheless, considering census features separately is justified well in terms of processing speed because they can be calculated much faster for the whole image.

The distinction between pixel-based gradient features, census features, and region-based structure features is important for real-time requirements. Pixel-based features are calculated beforehand for the whole image and reused when sliding the analysis window over the image. Region-based features have to be calculated separately for each analysis window. Pixel-based features are suited for fast candidate search, whereas more significant region-based features improve the performance of candidate verification. We choose a model size of 24 × 24 pixels that is commonly used for human face detection and obtain 484 gradient features, 484 census features, and 8,464 structure features. The first stages offer a quick candidate search and the final stages provide a more accurate but slower classification. The training procedure starts with randomly chosen nonface data for the initial stage. Following stages are trained with nonface data that are gathered by bootstrapping the model on images without ape faces. More details about the training procedure can be found in our previous work [9].

A 3 × 3 mean filter reduces noise in the input image. We resize the filtered image with different scaling factors and generate an image pyramid to detect faces of arbitrary size. The detection model of size 24 × 24 pixels analyzes each pyramid level with a coarse to fine search to further improve speed. Therefore, the detection model is shifted with a step size of about 6 pixel across each pyramid level. The neighborhood of a grid point is scanned more thoroughly only if the grid point produced a high face correlation score.

After the face detection process, we apply a subsequent eye search in all detected face regions with the same algorithms. We trained a detection model for each eye with a reduced size of 16 × 16 pixels. Only the eye regions were cut out from the annotated training data for this purpose. The eye models are simpler and less powerful compared to the face model and comprise five stages only and less features, because searching within face regions will lead to few false positives. Selected areas in all face regions around the left and right eye are scanned with the appropriate eye model in different scaling levels. Fixed eye markers of the face model are used if an eye could not be detected by the eye search.

3.2 Face alignment and lighting normalization

A very crucial step to achieve good performance in the subsequent face recognition task is the alignment of the detected faces. Based on the automatically detected eye coordinates, we first rotate the facial image into an upright position such that the eyes lie on a horizontal line. If coordinates for the center of the mouth are not available, we estimate the locations of the left and right corner of the mouth based on the eye coordinates only. However, if the location of the center of the mouth is provided, we can calculate the position of the mouth's corners more precisely, which will lead to a better alignment. Based on the locations of the left and right eye as well as the left and right corners of the mouth, we are then able to apply a projective transform to finally align the ape's face. This step ensures that facial features like eyes, nose, and mouth are located nearly at the same coordinates throughout the entire dataset. Consequently, this guarantees that extracted visual facial features are comparable for all faces. Figure 3 illustrates the applied face alignment procedure for an example image. After converting the aligned face image into gray scale, we apply a simple histogram equalization for lighting normalization and contrast adjustment.

3.3 Individual identification

The individual identification is the main part of the proposed system and consists of three steps: feature extraction, feature space transformation, and classification. In the first step we extract global as well as local visual features that are both well suited for discrimination. As those descriptors are too high dimensional to perform fast and robust face recognition in practice, we apply a feature space transformation technique called LPP [23] to achieve a lower dimensional subspace with only little loss of information that is important for identification. These lower dimensional feature vectors are then used for classification. After classifying the global and local feature vectors separately, we apply a decision fusion technique to get the final result.

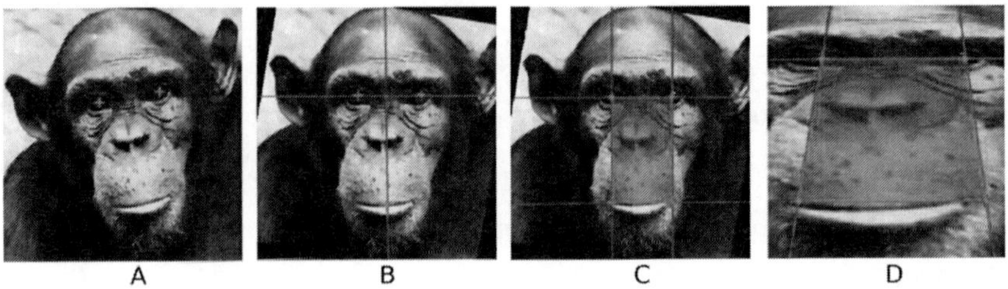

Figure 3. Face alignment. The face alignment procedure for an example image. Based on the detected eye coordinates the position of the mouth is estimated **(A)**. After rotating the facial image into an upright position**(B)**, such that both eyes lie on a horizontal line, the left and the right corner of the mouth is estimated **(C)**. Based on these four points a projective transformation is applied **(D)**. This ensures that facial features like eyes, nose, and mouth are located approximately at the same positions throughout the entire dataset, which is a prerequisite for accurate identification.

3.3.1 Feature extraction

Since global features gather holistic information of the face and local descriptors around facial points represent intrinsic factors, both should be used for classification. Additionally, it has been reported in the literature that different representations misclassify different patterns [35]. Therefore, various features offer complementary information which can be used to improve the recognition results. As for global features we propose to use Gabor features, which are known to perform well in pattern recognition tasks. The complimentary local descriptor is SURF, a powerful visual descriptor of interest points in an image.

Gabor descriptor

Gabor features are known to perform well in face recognition and pattern recognition tasks in general [36]- [38]. They are extracted by convolving the gray-level input image $I(z)$ with a set of Gabor kernels $\psi_{\mu,\nu}(z)$ as

$$G_{\mu,\nu}(z) = I(z) * \psi_{\mu,\nu}(z), \quad (1)$$

where $G_{\mu,\nu}(z)$ is the output image at orientation μ and scale ν at pixel location $z = (x,y)$. Complex Gabor kernels are defined as

$$\psi_{\mu,\nu}(z) = \frac{\|k_{\mu,\nu}\|^2}{\sigma^2} e^{\frac{-\|k_{\mu,\nu}\|^2\|z\|^2}{2\sigma^2}} [\, e^{ik_{\mu,\nu}z} - e^{\frac{-\sigma^2}{2}} \,], \quad (2)$$

where the wave vector $k_{\mu,v}$ is defined as $k_{\mu,v} = k_v e^{i\theta_\mu}$ with $k_v = \frac{k_{\max}}{f^v}$ and $\theta_\mu = \frac{\pi\mu}{8}$. The maximum frequency is denoted as $k_{\max}$ and f is the spacing between kernels in the frequency domain. Furthermore, σ represents the ratio of the Gaussian window to the wavelength.

In general, $G_{\mu,v}(z)$ is complex and can be rewritten as $G_{\mu,v}(z) = M_{\mu,v}(z)e^{i\vartheta_{\mu,v}(z)}$, where $M_{\mu,v}(z)$ denotes the magnitude, and $\vartheta_{\mu,v}(z)$ the phase at pixel location z. Since the magnitude contains the local energy variation in the facial image, $M_{\mu,v}$ is used as feature, while $\vartheta_{\mu,v}(z)$ is ignored for further processing. Finally, the overall feature vector is constructed as

$$x_{\text{GABOR}} = \left(m_{0,0}^{(\rho)}, m_{0,1}^{(\rho)}, \cdots, m_{1,0}^{(\rho)}, \cdots, m_{K,L}^{(\rho)}\right), \tag{3}$$

where $m_{\mu,v}^{(\rho)}$ is a column vector representing the normalized and vectorized version of the magnitude matrix $M_{\mu,v}$ which was down sampled by factor ρ.

For feature extraction we use five scales and eight orientations for the generation of Gabor kernels with size of 31×31. We chose to set $k_{\max} = \frac{\pi}{2}, f = \sqrt{2},$ and $\sigma = \pi$. After convolving an image with the resulting 40 Gabor wavelets, we down-sample the magnitude matrix $M_{\mu,v}$ by a factor of $\rho = 8$ by using a bilinear interpolation.

SURF descriptor SURF is a fast and robust scale- and rotation-invariant interest point detector and descriptor. It was first published by Bay et al. in 2008 [10]. In this task we already know the position of the interest points so that we only refer to the descriptor part of SURF in this paper. In the following we briefly describe the main ideas of SURF. A more detailed description including the detection of interest points can be found in [10]. As claimed by the authors, the standard version of SURF is several times faster, more compact, and, at the same time, more robust against certain image transformations than comparable local descriptors like scale invariant feature transform (SIFT) [39]. Similar to SIFT and its variants, SURF describes the distribution of intensity content within a certain neighborhood around the interest point. However, instead of using gradient information directly, SURF uses first-order Haar wavelet responses in x and y-direction. For efficiency, SURF exploits integral images which drastically reduces processing time while at the same time improves the robustness of the resulting descriptor. In order to increase the robustness against rotation, usually the first step of feature extraction is to identify a reproducible orientation for the interest point. The dominant orientation can be found by calculating the sum of the Gaussian-weighted Haar wavelet responses using a sliding window around a circular region around the interest point. The next step is to construct a square region with correct orientation symmetrically around the interest point. This region is then split into 4×4 subregions. Finally, the feature vector can be calculated by again using Haar wavelet responses weighted with a Gaussian kernel which is centered at the particular interest point. The horizontal and vertical wavelet responses, *dx*and *dy*, as well as their absolute values are summed up over each subregion to construct the final feature vector of size 64

$$x_{\text{SURF}} = \left(\sum dx, \sum dy, \sum |dx|, \sum |dy|\right). \tag{4}$$

We extract SURF descriptors on six facial fiducial points which are calculated based on the detected eye markings. Figure 4 shows the location of the facial markings. Note that for the local feature extraction, we just rotate the faces into an upright position for alignment instead of applying a projective transform to prevent unnatural distortion of local regions and then resize the facial image to 64×64 pixels. Since we already performed this step during face alignment as discussed in Section 3.2, we do not use the rotation-invariant version of the SURF descriptor. This saves computation time because it is not necessary anymore to identify the main orientation of the interest point beforehand. As stated in [10], the upright version of SURF is faster to compute and can increase distinctiveness while maintaining a certain robustness against small rotation angles of up to ±15°. Furthermore, we only need to compute the descriptor in one particular scale because we previously resized every face image to a fixed size. This also makes the feature extraction step more efficient and facilitates real-time performance of the final system. Based on the assumption that wrinkle patterns under and between the eyes are unique across individuals and useful for identification, the first three points are located under the left and right eye, as well as between both eyes. Furthermore, we assume that the area around the nose is well suited for discrimination. Therefore, we use the tip of the nose as well as the left and the right nostril as additional locations for local feature extraction. We do not extract information out of the mouth region because we noticed that this area is often subject to occlusion and deformation because of eating and facial expressions. Extracting features out of this region would lead to a high intra-class variance and would hamper classification.

3.3.2 Feature space transformation

The goal of many feature space transformation techniques is to project the *N* high-dimensional feature vectors $\{x_1,\cdots,x_N\}$ of size *n* into a smaller dimensional subspace of size *m* using a unitary projection matrix $W \in \mathbb{R}^{n\times m}$

$$y_k = W^T x_k; \quad \text{with} \quad x_k \in \mathbb{R}^{n \times 1}, y_k \in \mathbb{R}^{m \times 1}, m \leq n. \tag{5}$$

Figure 4. Local feature extraction. The positions of the applied SURF descriptor for local feature extraction. Three out of six positions for local feature extraction are located under and between both eyes; the remaining three interest points are situated on the nose tip and both nostrils. The mouth region is not used for feature extraction because we especially noticed that this region is often subject to occlusion and facial expressions.

The resulting feature vectors $yk \in R^{m \times 1}$, with $k = 1, \cdots, N$, can then be used for classification.

LPP [23] assumes that the feature vectors reside on a nonlinear submanifold hidden in the original feature space. LPP tries to find an embedding that preserves local information by modeling the manifold structure of the feature space using a nearest-neighbor graph. First, an adjacency graph G is defined. An edge is put between two nodes $k$and j if they belong to the same class C.

LPP will try to optimally preserve this graph when choosing projections. After constructing the graph, weights have to be assigned to the edges. Therefore, a sparse symmetric matrix S of size $N \times N$ is created with $S_{k,j}$ having the weight of the edge joining vertices k and j, and 0 otherwise. The weights are calculated as follows:

$$S_{k,j} = \begin{cases} e^{-\frac{\|x_k - x_j\|^2}{2*\sigma^2}}, & \text{if} \quad C(x_k) = C(x_j) \\ 0, & \text{otherwise.} \end{cases} \tag{6}$$

Here, σ denotes a constant for normalization which we set to 100 in our system. The objective function of LPP is defined as

$$w_{\text{opt}} = \min \sum_{kj} (y_k - y_j)^2 S_{k,j}. \tag{7}$$

Following some simple algebraic steps, it is possible to show that Equation 7 finally results in a generalized eigenvalue problem:

$$XLX^T w = \lambda XDX^T w, \tag{8}$$

where D is a diagonal matrix whose entries are column sums of S and $L = D - S$ is the so-called Laplacian matrix. The k th column of matrix X is x_k.

The projection matrix W is constructed by concatenating the solution to the above equation, i.e., the column vectors of $W_{LPP} = [w_1, \cdots, w_m]$ are ordered ascendingly according to their eigenvalues. Usually, the original features are first projected into the PCA subspace before applying LPP by deleting the smallest principle components. Thus, the final embedding is as follows:

$$W_{\text{final}} = W_{\text{PCA}} W_{\text{LPP}}. \tag{9}$$

Details about the algorithm and the underlying theory can be found in [23].

3.3.3 Classification

Sparse representation classification

For the classification of global features, we use the SRC paradigm developed by Wright et al., which is known to perform well for face recognition [24, 25].

Let $\tilde{A} \in R^{m\times l}$ be the normalized matrix of training samples transformed into the feature space and t˜∈Rm×1$\tilde{t} \in \mathbb{R}^{m\times 1}$ be the normalized transformed feature vector of the test image, where *m* is the dimensionality of the feature space and *l* the number of training samples. Classification can be done by first solving a convex optimization problem via l_1-norm minimization:

$$\hat{p} = \arg\min_{p} \|p\|_1 \quad \text{subject to} \quad \tilde{t} = \tilde{A}p, \tag{10}$$

where $p \in R^{l\times 1}$ is a sparse coefficient vector whose entries only associated with the *i* th class should be 1 and the rest be 0. In other words, we try to represent the feature vector t˜ of the test image as a linear combination of the training samples of the same class. Therefore, ˜$\tilde{t}$ is assigned to the object class that minimizes the residual ri(t˜) between t˜ and $\tilde{A}$(δi⊙p^) such that

$$\min_{i} r_i(\tilde{t}) = \|\tilde{t} - \tilde{A}(\delta_i \odot \hat{p})\|_2, \tag{11}$$

where ⊙ denotes the elementwise multiplication known as Hadamard product. The vector δi∈Rl×1$\delta_i \in \mathbb{R}^{l\times 1}$ is called the characteristic function of class *i*. It is a filter vector which is 1 for all training samples of class *i* and 0 elsewhere. A detailed description of SRC can be found in [24].

Support vector machines

In the proposed system, we use a support vector machine (SVM) [40] for the classification of local features. SVM is a discriminative classifier, attempting to generate an optimal decision plane between feature vectors of the training classes. Often, the classification with linear separation planes is not possible in the original feature space for real-world applications. Using a so-called kernel trick, the feature vectors are transformed into a higher dimensional space in which they can be linearly separated. We use a radial basis function (RBF) as kernel in this paper.

3.3.4 Decision fusion

The decision fusion paradigm we use in this paper was influenced by ideas of [41]. A parallel ensemble classifier which fuses the rank outputs of different classifiers is used to combine the results of local and global features. In contrast to the parallel fusion scheme proposed in [41], where only a single weighting function $w(\mathcal{R}) = \mathcal{R}^c$ for rank $\mathcal{R}$ and constant *c* is used as nonlinear rank sum method, we weight the results of both classifiers using different weighting functions for every classifier. Additionally, the confidences of each classifier can be taken into account when generating the weighting function $w(\mathcal{R}) = e^{s(\mathcal{R})}$, where $s(\mathcal{R})$ represents the confidence of SRC or SVM for rank $\mathcal{R}$. For SRC we use the vector of residuals from Equation 11 as confidence measure, while for SVM the probability estimates of LibSVM [40] can be utilized. The probability estimates can simply be converted into match scores by negating the probabilities. Details on the estimation of probabilities for SVM can be found in [42]. The final score vector $s_f \in \mathbb{R}^{C \times 1}$, where *C* is the number of classes, is then simply the sum of both weighting functions: $s_f = w_{SRC} + w_{SVM}$. Finally, s_f is ordered ascendingly to obtain the final result.

Figure 5 illustrates the parallel fusion scheme we use in this paper. Note that for every of the six facial interest points, we transform the resulting SURF descriptors separately into a smaller dimensional subspace before concatenating them to get the final local feature vector.

4 EXPERIMENTS AND RESULTS

4.1 Dataset description

Due to the lack of publicly available face databases of chimpanzees, we use self-assembled annotated datasets of captive as well as free-living individuals from the Zoo Leipzig, Germany, and the Taï National Park, Côte d'Ivoire, Africa, respectively. For benchmark purposes, the license rights for both datasets can be purchased over our project website http://www.saisbeco.com. Both datasets were annotated by experts. Table 1 gives details about both datasets that were used in our experiments.

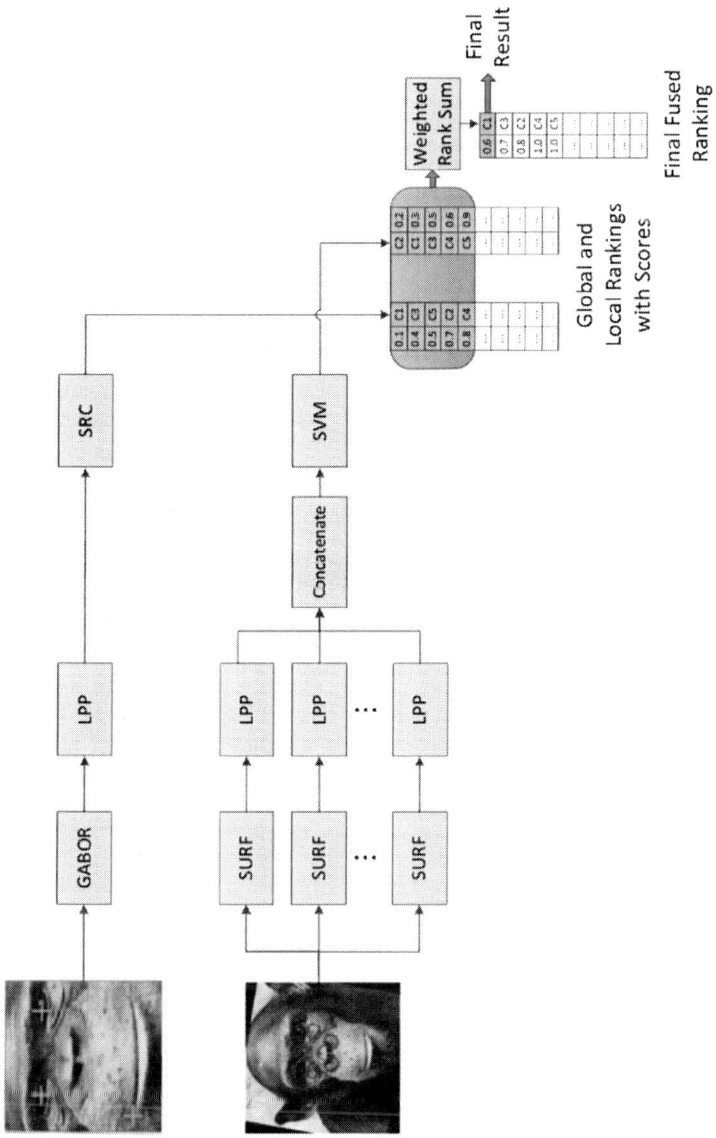

Figure 5. Proposed parallel fusion scheme. This figure shows the parallel fusion scheme to combine the results of global Gabor features and local SURF descriptors. Both global and local features are first projected into a smaller dimensional subspace using LPP. Note that we transform each SURF feature separately into the feature space before concatenating the resulting vectors to a comprehensive local feature vector. The global feature is classified using SRC while the local feature vector is classified by SVM with RBF kernel. The ranked results are then combined using the decision fusion rank sum method explained in Section 3.3.4 to obtain the final result.

Table 1. Overview of the datasets we used in our experiments

Dataset	Images	Individuals	Faces	$\sum$ Pixels (MP)
ChimpZoo	2,617	24	598	6,403
ChimpTaï	3,905	71	1,432	5,409

Example images for both datasets with detected faces can be seen in Figure 6. To have a valid ground truth for evaluation, the position of the head, eyes, and mouth was annotated. Metadata were also assigned to every annotated face such as gender, age and the name of the individual. The experts used our proprietary annotation tool for this purpose. This tool allows the annotation of face regions in the image along with related facial marker points. Moreover, metadata can be assigned to all faces by additional attributes. The annotations are stored separately for each image in a XML file. More details about the annotation tool can be found in our previous work [43]. Figure 7shows detected faces of one individual for the ChimZoo dataset (Figure 7A) and the ChimpTaï dataset (Figure 7B), respectively. It is obvious that both datasets are very challenging for the recognition task because detected faces of one single individual can have a variety of poses, expressions, lighting situations and even partial occlusion by branches or leafs. Thus, the algorithm used for identification is required to be robust against that kind of variations to achieve sufficient recognition results.

A B

Figure 6. Examples of detected faces. Example images of detected faces for the ChimpZoo dataset **(A)** and the ChimpTaï dataset **(B)**. The region of the successfully detected faces and eyes are marked (in green lines). Additionally, the species is automatically assigned to the face.

A

B

Figure 7. Detected faces of one individual per dataset. Detected faces of one individual of the ChimpZoo dataset**(A)** and the ChimpTaï dataset **(B)**. Both datasets are very challenging due to difficult lighting situations, facial expressions, poses, and even partial occlusion by branches or leafs.

4.2 Evaluation measures and experiment design

Since the face detection stage will produce false-positive detections, we decided to use an open-set identification scheme to deal with that issue. We use the performance statistics described in [44]- [46] to evaluate our system. In open-set identification, first the system has to decide if the probe p_j represents a sample of an individual in the gallery $\mathcal{G}$ or not. If the system decided that the individual in the probe is known to the system, then it also has to report the identity of the individual. While for a closed-set identification, the question is how many test images are correctly classified as a certain individual, two more types of errors can occur for an open-set classification. Additional to false classifications, it is also possible that the system rejects known individuals or accepts impostors. Let $\mathcal{P_G}$ be the probe set that contains face images of chimpanzees in the gallery and $\mathcal{P_N}$ the probe set that contains samples of chimpanzees that are not known to the system. When a probe p_j is presented to the system, a score vector $s \in \mathbb{R}^{C\times 1}$ can be calculated, where C is the number of known individuals in the database. The entries of this vector are scaled between 0 and 1. The smaller the value, the higher the confidence of the classifier. For SRC we use the vector of residuals r from Equation 11 as confidence measures, while for the proposed decision fusion technique, the combined weightings s_f can be used as score values for each class. For classification by SVM, the probabilities of the classifier can be negated to assign them to the score vector s_f.

A probe p_j is detected and identified if the minimal score $s_{\min,j}$ is below the operating threshold τ and identified correctly with rank $(p_j) = 1$. Therefore, the detection and identification rate P_{DI} at threshold τ can be calculated as

$$P_{\mathrm{DI}}(\tau, 1) = \frac{\left|\{p_j : p_j \in \mathcal{P_G}, \mathrm{rank}(p_j) = 1, \text{and } s_{\min,j} \leq \tau\}\right|}{|\mathcal{P_G}|}. \tag{12}$$

The second performance measure is the false alarm rate P_{FA}. A false alarm occurs when the minimal match score of an impostor is below the operating threshold τ. Consequently, the false alarm rate is the fraction of probes in $\mathcal{P_N}$ that are detected as genuine individuals and is calculated as

$$P_{\mathrm{FA}}(\tau) = \frac{\left|\{p_j : p_j \in \mathcal{P_N}, s_{\min,j} \leq \tau\}\right|}{|\mathcal{P_N}|}. \tag{13}$$

An ideal system would have a detection and identification rate of 1.0 and a false alarm rate of 0.0, which means that all individuals are detected and classified correctly and there are no false alarms. In practice however, both measures have to be traded-off against each other. This trade-off is shown in a receiver-operating characteristic (ROC) by iteratively changing the operating threshold τ. Another important performance statistic is the equal error rate (EER). It is reached when the false alarm rate is equal to the false detection and identification rate $P_{FA} = 1 - P_{DI}$.

In addition to false-positive detections, one individual at a time is removed from the training set and presented it as an impostor to test the system's capability to reject unknown chimpanzees. This procedure is repeated C times, where C is the number of individuals in the dataset, such that every chimp takes the role of an impostor once. To get valid results, we additionally apply a tenfold stratified cross validation. Images of false-positive detections as well as all pictures of the unknown individual remain in the test set for all ten folds and are not used for training. We only consider detections with a minimum size of 64×64 pixels for identification, which dramatically decreases the number of false-positive detections. Furthermore, we only focus on individuals with at least five detected face images in the database to get an appropriate number of training images for each class. This limitation results in 24 individuals for the ChimpZoo and 48 subjects for ChimpTaï dataset. After aligning the detected face images as described in Section 3.2, we apply a histogram equalization for lighting normalization. To make the results comparable, we chose to have a feature dimension of 160 for all applied feature space transformation techniques. For the local SURF features, we transform the resulting feature vectors separately into a smaller dimensional subspace of size 50 for every of the six used facial fiducial points before concatenating them to the final feature vector. This results in a local feature vector of size 6 × 50 = 300.

4.3 System evaluation

4.3.1 Face detection

Because of the lack of publicly available face databases for chimpanzees, we trained the detection model with frontal faces of the ChimpZoo dataset and evaluated it on both datasets. Figure 8 shows the results with ROC curves. The detection rate on the ChimpZoo dataset is considerably higher. This can be expected because this dataset was used for training. Moreover, it shows a higher image quality than the ChimpTaï dataset in terms of resolution and extrinsic factors like lighting conditions, contrast, and occlusion. A threshold defines the working point of the detector on the ROC curve. If we accept 0.1 false positives per megapixel in practice, the detector finds 93% and 82% of the faces in the ChimpZoo and ChimpTaï dataset, respectively.

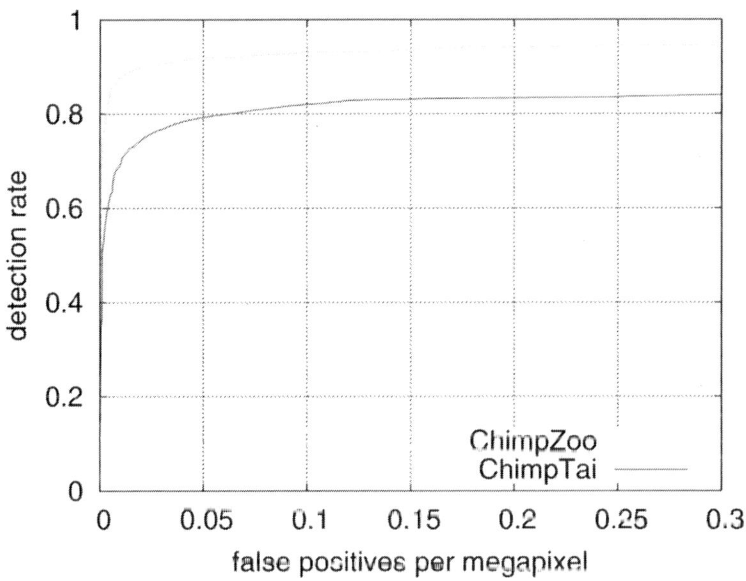

Figure 8. Face detection performance. This figure shows the ROC curve of the face detection model evaluated on the ChimpZoo and ChimpTaï dataset. The number of false detections is normalized to the total pixel sum of both datasets.

4.3.2 Face identification

Experiment 1: influence of visual features and feature space transformation

In the first experiment we want to address the question if Gabor features (GABOR) or pixel based features (PIXEL), used in our previous work [5, 7], are better suited for face recognition of great apes. Furthermore, we evaluate and compare three different feature space transformation techniques: random projection (RAND) [24], PCA [21], and LPP [23]. To make the results comparable, we set the number of features to 160 for every feature space transformation method. For the face alignment procedure, we used the manually annotated facial marker points in this experiment. Figure 9 shows the results for the ChimpZoo dataset (Figure 9A) and ChimpTaï dataset (Figure9B), respectively. The black diagonal line denotes the line of equal error. For all combinations we used the SRC algorithm for classification. As can be seen, our approach to use Gabor features as global descriptors and LPP for feature space transformation outperforms all the other approaches on both datasets (blue solid line). The EER for every algorithm and both datasets can be seen in Table 2.

Since global Gabor features in conjunction with LPP achieves the best results in the first experiment, this combination should be used for holistic face recognition for primates. However, in the next experiment we will show that this algorithm can still be enhanced by additionally using locally extracted SURF features and our proposed decision-based fusion scheme.

Table 2. EER for Gabor and pixel-based features for feature space transformation

	ChimpZoo		ChimpTaï	
EER	**Pixel**	**Gabor**	**Pixel**	**Gabor**
RAND	0.3804	0.2885	0.5171	0.4504
PCA	0.2671	0.2161	0.4363	0.3647
LPP	0.2313	**0.1290**	0.3950	**0.2938**

Equal error rates (EER) for Gabor and pixel-based features in combination with random projection (RAND), principal component analysis (PCA), and locality preserving projections (LPP) for feature space transformation. Our proposed approach is in boldface and performs best on both datasets.

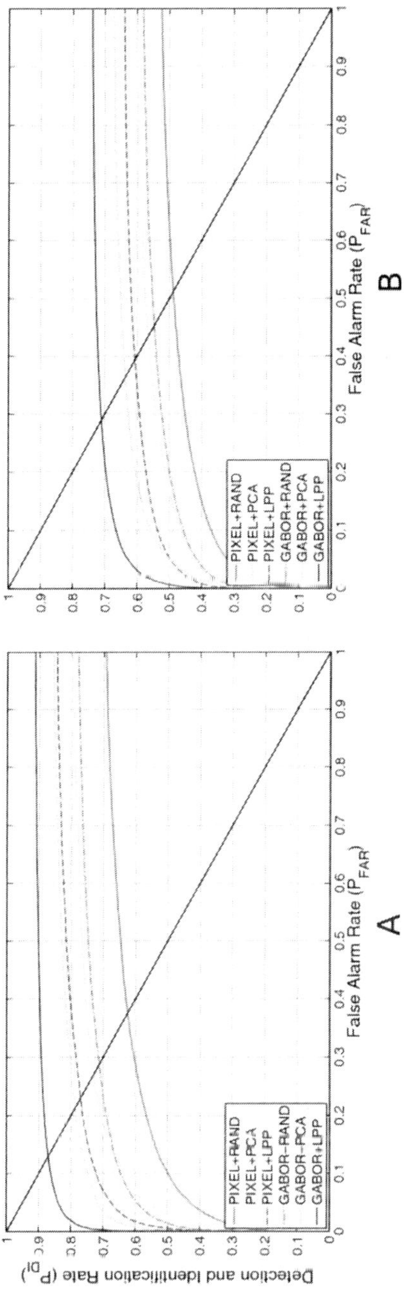

Figure 9. Results of experiment 1. ROC curves for the first experiment we conducted in this paper for the ChimpZoo dataset **(A)** and the ChimpTaï dataset **(B)**. The black solid line denotes the line of equal error. We compared globally extracted Gabor features (GABOR) with pixel-based features (PIXEL).

We combined the features with three different methods for feature space transformation, random projection (RAND), Principal component analysis (PCA), and locality preserving projections (LPP). For all combinations we used the SRC algorithm for classification. It can be seen that Gabor features perform best in most of the cases and are therefore better suited for describing chimpanzee faces than simple pixel-based features. Our proposed approach (GABOR + LPP), which is denoted by the solid blue line, outperforms all the other algorithms with an equal error rate (EER) of 0.1290 and 0.2938 for the ChimpZoo and ChimpTaï dataset, respectively.

Experiment 2: combination of global and local features

In the second experiment, we show that our proposed approach for combination of global and local features improves the performance and outperforms systems based on global or local features alone. As specified in Section 3, we use Gabor features as global face representation and SURF as local descriptors. Both features are transformed into a lower dimensional subspace using LPP. Note that we apply LPP on SURF features separately for every interest point before generating the final local feature vector. Again, for the global Gabor feature vector, we set the number of dimensions to 160 while each SURF descriptor is transformed into a feature space of size 50, resulting in a combined local descriptor of size 300. The classification is done separately for global and local features using SRC and SVM, respectively. The results for global and local features are combined in the decision-based manner we described in Section 3.3.4. Like in experiment 1, the manually annotated eye coordinates were used for alignment. Figure 10 shows the resulting ROC curves for the ChimpZoo and ChimpTaï dataset, respectively. For both datasets the global Gabor features (green dash-dotted line) perform significantly better than the local SURF descriptors (red dashed line). Nevertheless, obviously SURF descriptors encode important information to discriminate between individuals which can be seen from the results of the proposed fusion paradigm (blue solid line). The decision fusion of global and local features performs better than global and local features alone, especially for the free-living individuals (Figure 10B). The associated equal error rates can be seen in Table 3.

It is obvious that our proposed fusion scheme performs better than global and local features alone. Therefore, the idea of using the confidences of both classifiers improves the performance of the face recognition algorithm for chimpanzee faces in real-world environments.

However, we still used manually annotated eye coordinates for alignment and estimation of facial fiducial points for local feature extraction. In the final experiment we use the automatically detected facial markings for this purpose.

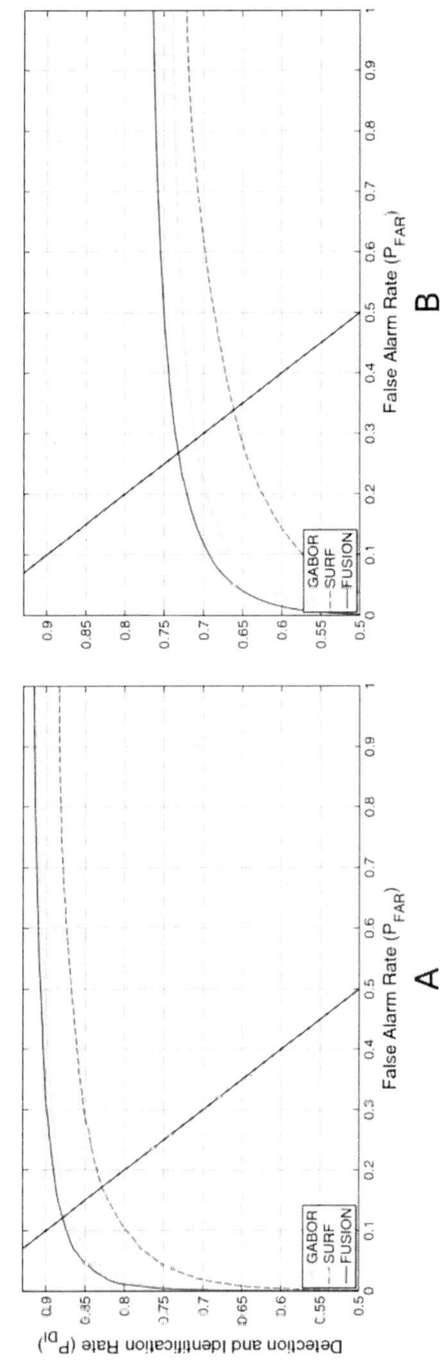

Figure 10. Results of experiment 2. Achieved results for the second experiment we conducted for the ChimpZoo dataset **(A)** and the ChimpTaï dataset **(B)**.

Again, the black solid line denotes the line of equal error. The green dash-dotted line shows the result for globally extracted Gabor features in combination with LPP and SRC. The red dashed line denotes the results for the local SURF descriptors extracted around six facial feature points. Each feature vector was transformed separately into a smaller dimensional subspace using LPP before classifying the concatenated final feature vector using a SVM with an RBF kernel. Although the locally extracted features perform worse than the global descriptor, they carry important information for classification. This can be seen from the results of the decision fusion scheme (blue solid line) that outperforms algorithms based on global and local features alone with an EER of 0.1189 and 0.2669 for the ChimpZoo and the ChimpTaï dataset, respectively.

Table 3. EER for global Gabor and local SURF features and proposed parallel decision fusion scheme

EER	Gabor	SURF	Fusion
ChimZoo	0.1290	0.1903	**0.1189**
ChimpTaï	0.2938	0.3171	**0.2669**

The proposed decision fusion method is printed in bold and performs best on both datasets.

Experiment 3: manually annotated vs. automatically detected facial markings

In the third and last experiment, we show that automatically detected facial feature points for face alignment perform almost as good as manually annotated ones. For facial feature detection, we use the algorithm described in Section 3.1. For face identification we use the proposed system that combines global and local features for recognition which performed best in experiment 2. All parameters were set as described in the previous experiment. We compare the recognition results of the system for face alignment with manually annotated feature points, automatically detected facial feature points, and if no alignment was applied at all. Figure 11 shows the ROC curves of the proposed identification algorithm for alignment using the manually annotated facial feature points (blue solid line), automatically detected markings (red dashed line), and without applied alignment (green dash-dotted line) for the ChimpZoo dataset in Figure 11A and the ChimpTaï dataset in Figure 11B. The according equal error rates can be seen in Table 4.

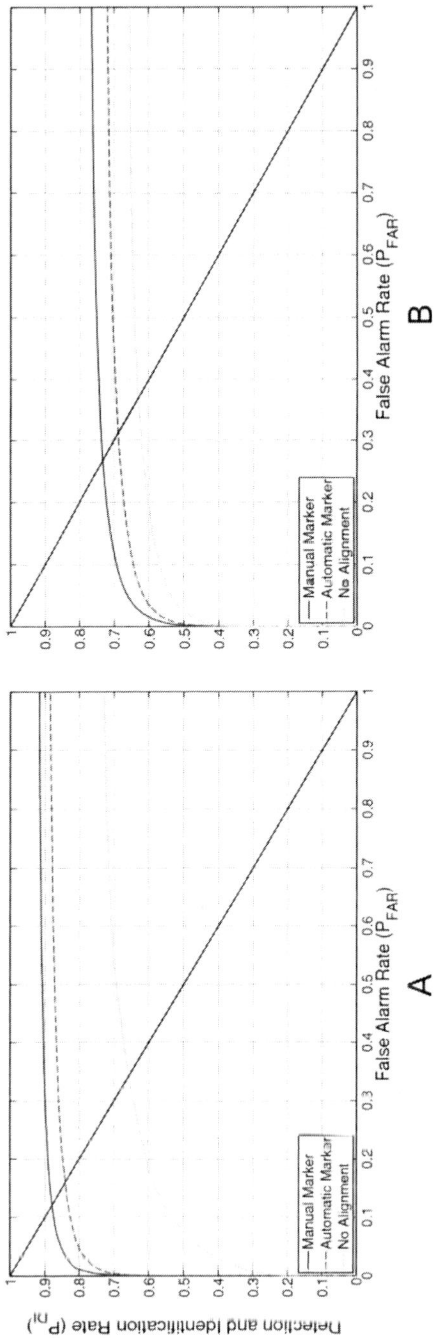

Figure 11. Results of experiment 3. ROC curves of the proposed identification system for alignment using manually annotated coordinates of both eyes and

mouth (blue solid line), automatically detected eyes (red dashed line), and without applied alignment (green dash-dotted line). **(A)** Curves for the ChimpZoo dataset. **(B)**Results for the ChimpTaï dataset. The black solid line denotes the line of equal error. As can be seen, the performance of the algorithm if manually annotated facial markings were used for alignment is only slightly better than alignment using automatically detected facial feature points. The accuracy of the system drops significantly if no alignment was applied at all.

Table 4. EER of the proposed identification algorithm if alignment was applied

EER	Manual markings	Automatic markings	No alignment
ChimZoo	0.1189	0.1590	0.3447
ChimpTaï	0.2669	0.3103	0.3668

EER of the proposed identification algorithm if alignment was applied using manually annotated markings, automatically detect markings, or if no alignment was applied as a pre-processing step. The EER for manually annotated markings and automatically detected feature points are very close while the performance of the algorithm drops significantly if no alignment was performed at all.

If manual markings were used for alignment and estimation of the facial fiducial points for local feature extraction, the proposed algorithm performs best. However, if we use automatically detected eye coordinates, the performance of the algorithms is only slightly worse than for manually annotated markings. This is because the automatic detection of eye coordinates is not always as accurate as manually detected ones. Another explanation is that for the automatically detected markings, it was only possible to estimate the coordinates for local feature extraction based on the location of both eyes. For the manually annotated ones, however, we additionally used the annotated location of the mouth to estimate these locations more precisely. Therefore, the local feature extraction is much more accurate if an exact location of the mouth region is available.

In the previous three experiments, we showed that the face recognition algorithm proposed in this paper achieved excellent results and outperformed the approaches of previous works. However, we only showed the relationship between correct detection and identification rate (P_{DI}) and percentage of impostors accepted by the system (P_{FAR}). Moreover, another important question is how the system's overall error rate is influenced by the other two types of errors, the false rejection and the false classification. This issue is depicted in Figure 12, showing the results of the proposed system using automatically detected facial markings for ChimpZoo (Figure 12A) and ChimpTaï (Figure 12B). The blue area denotes the rate of false rejections (P_{FR}), while the red area shows the influence of false classification (P_{FC}) for different false alarm rates (P_{FA}). The lower bound depicts the ROC curve from Figure 11 (red dashed line). The false classification rates and false rejection rates for both datasets at the point of equal error can be seen in Table 5.

It can be seen that for the ChimpZoo dataset, the main contribution to the overall error rate of the system is caused by falsely rejected faces of genuine individuals with P_{FR} of 12.53%. Only 3.52% was due to false classifications. This shows that many facial images of known identities were rejected as impostors because of too much pose variation, facial expressions, or occlusions. For the ChimpTaï dataset, however, the system's performance is almost equally caused by false classification, with P_{FC} of 17.73% and P_{FR} of 13.52%. This shows that the ChimpTaï dataset is much more challenging than the ChimpZoo dataset because it was gathered in a wildlife environment. Furthermore, the ChimpTaï dataset contains twice as much individuals at a much lower quality which again explains the strong influence of false classifications to the overall error of the proposed system.

5 CONCLUSIONS

In the ongoing biodiversity crisis, many species including great apes like chimpanzees for instance are threatened and need to be protected. An essential part of efficient biodiversity and wildlife conversation management is population monitoring and individual identification to estimate population sizes, asses viability, and evaluate the success of implemented protection schemes. Therefore, the development of new monitoring techniques using autonomous recording devices is currently of intense research [47]. However, manually processing large amounts of data is a tedious work and therefore extremely time-consuming and highly cost-intensive.

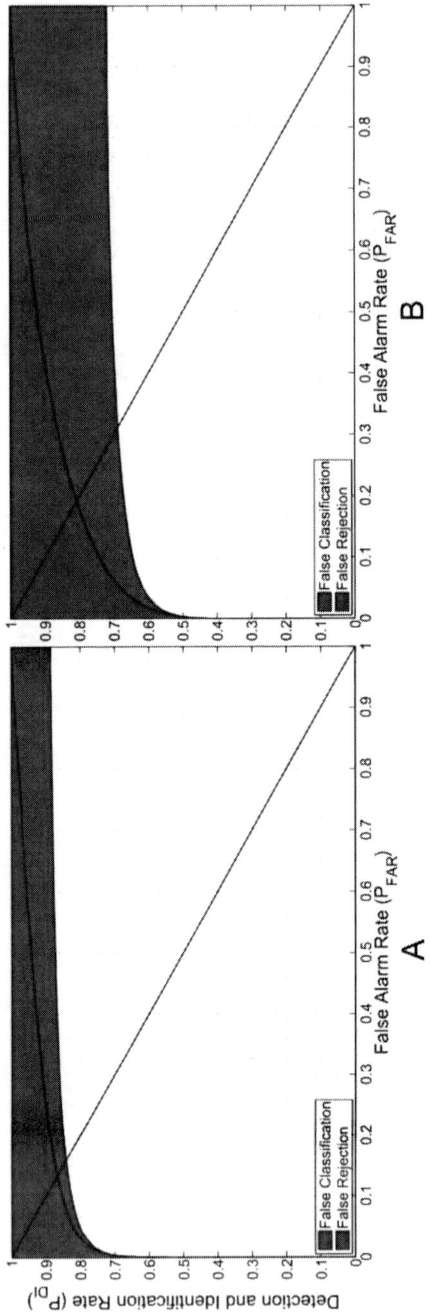

Figure 12. Contribution of P_{FR} and P_{FC} to the overall accuracy of the system. This figure depicts the contribution of the false rejection rate (P_{FR}) and the false classification rate (P_{FC}) to the overall performance of the proposed

system when automatically detected facial features were used for alignment and estimation of other locations for local feature extraction. **(A)** Curves for the ChimpZoo dataset. **(B)** Results for the ChimpTaï dataset. The black solid line denotes the line of equal error. The blue area represents the influence of the false rejections, while the red area shows the contribution of the false classifications. For the ChimpZoo dataset, the error rate of the system at the point of equal error is mainly caused by false rejection of genuine individuals, and only 3.52% is due to false classifications. For the ChimpTaï dataset, however, 13.52% of the overall error rate is caused by false rejections, while 17.72% are caused by false classifications.

Table 5. P_{FR} and P_{FC} at EER for both datasets

	P_{FR} at EER	P_{FC} at EER
ChimZoo	0.1253	0.0352
ChimpTaï	0.1352	0.1773

False rejection rate (P_{FR}) and false classification rate (P_{FC}) at the point of equal error (EER) for both datasets, ChimpZoo and ChimpTaï, respectively.

To overcome these issues, we presented an automated identification framework for chimpanzees in real-world environments in this paper. Based on the assumption that humans and chimpanzees share similar properties of the face, we proposed to use the face detection and recognition technology for identification of great apes in our previous work [5]- [9]. In this paper we successfully combined face detection, face alignment, and face recognition to a complete identification system for chimpanzee faces in real-world environments. We successfully combined globally extracted holistic features and local descriptors for identification using a decision fusion scheme. As global features we used the well-established Gabor features. We transformed the resulting high-dimensional feature vectors into a smaller, more discriminating subspace using LPP. For classification we used an algorithm called SRC. Since it is known from the literature that different features encode different information, we also extract SURF around local facial feature points to make the system more robust against difficult lighting situations, various poses and expressions as well as partial occlusion by branches, leafs, or other individuals. We separately transformed the resulting SURF descriptors into a lower dimensional subspace for every facial fiducial point. After concatenating the resulting low-dimensional descriptors to get one comprehensive vector of local features, we use SVM with RBF kernel for classification. We combine the classification results of global and local features in a decision-based manner by taking the confidences of both classifiers into account. Furthermore, we thoroughly evaluated our proposed algorithm on two datasets of captive and free-living chimpanzee individuals which were annotated by experts using an open-set classification scheme. In the three experiments we showed that our approach outperforms previously presented algorithms for chimpanzee identification. Although both datasets were gathered in real-world environments, opposed to most datasets used to evaluate algorithms for human face recognition, our system performs very well and achieves promising results. Therefore, the presented framework can be applied in real-life scenarios for identification of great apes. Thus, the system will assist biologists, researchers, and gamekeepers with tedious annotation work of gathered image and video material and therefore has the potential to open up new venues for efficient and innovative wildlife monitoring and biodiversity conservation management. Currently, intensive pilot studies using autonomous infrared-triggered remote video cameras are conducted in Loango National Park, Gabon [48] and Taï National Park, Côte d'Ivoire [49]. These studies have provided promising results in both number of species detected, as well as visitation rates, demonstrating the potential of such an approach for biomonitoring. Our proposed framework for automatic detection and identification of chimpanzees will help researchers to efficiently scan and retrieve video sequences that are important for biologists, i.e., where chimpanzees or other great apes are present. After providing an annotated dataset of labeled chimpanzee faces, the system will also be able to recognize known and reject unknown individuals. Although grouping similar-looking faces of unknown individuals remains a future work, such an approach could help biologists to expand the dataset of known chimpanzees over time and successively improve the accuracy of the system. Hence, biologists are then able to conduct biodiversity time series analysis to assess whether significant fluctuations in biodiversity occur.

Although the presented system achieved very good results on both datasets, we hope to further increase the performance of the system by extending the approach for face recognition in video. Because the temporal component of video can contain important information for identification, we expect further improvement of the system by exploiting temporal information. For example, finding the shots in a video sequence which are best suitable for face recognition in terms of pose, motion blur, and lighting could be one approach to extend the system towards video. Furthermore, frame weighting algorithms or techniques like super-resolution are conceivable to take advantage of temporal information in video sequences. In addition, automatic detection of more facial features could lead to better alignment and more precise localization of facial fiducial points for local feature extraction, which will further improve the performance of the system.

ACKNOWLEDGEMENTS

This work was funded by the German Federal Ministry of Education and Research (BMBF) under the 'Pact for research and innovation'. We thank the Ivorian authorities for the long-term support, especially the Ivorian Ministére de l'Environnement, des Eaux et Forêts and the Ministére de l'Enseignement Supérieur et de la Recherche Scientifique, the directorship of the Taï National Park, the OIPR, and the CSRS in Abidjan. Financial support is gratefully acknowledged from the Swiss Science Foundation. We would like to thank especially Dr. Tobias Deschner for collecting videos and pictures over the last years and for providing invaluable assistance during the data collection. We thank all the numerous field assistants and students for their work on the Taï Chimpanzee Project. We thank the Zoo Leipzig and the Wolfgang Köhler Primate Research Center (WKPRC), especially Josep Call and all the numerous research assistants, zoo-keepers, and Josefine Kalbitz for their support and collaboration. We also thank Laura Aporius for providing the videos and pictures in 2010. This work was supported by the Max Planck Society. We also thank Laura Aporius for the annotation of data.

COMPETING INTERESTS

Both authors declare that they have no competing interests.

REFERENCES

1. Hilton-Taylor C, Stuart SN: *Wildlife in a Changing World - an Analysis of the 2008 IUCN Red List of Threatened Species*. Gland Switzerland; IUCN 2009. http://www.iucnredlist.org/technical-documents/references

2. Walsh PD, Abernethy KAA, Bermejo M. Catastrophic ape decline in western equatorial Africa. *Nature* 2003, 422: 611-614.

3. Campbell G, Kuehl H, Kouame PN, Boesch C: Alarming decline of west African chimpanzees in Côte d'Ivoire.*Current Biology* 2008, 18(19):R904-905

4. Rowcliffe JM, Carbone C: Surveys using camera traps: are we looking to a brighter future? *Anim. Conserv* 2008, 11(3):185-186.

5. Loos A, Pfitzer M, Aporius L: Identification of great apes using face recognition. In *19th European Signal Processing Conference (EUSIPCO)*. Barcelona; 29 August–2 September 2011.

6. Loos A: Identification of great apes using Gabor features and locality preserving projections. In *1st ACM International Workshop on Multimedia Analysis for Ecological Data (MAED) in Conjunction with ACM Multimedia*. New York: ACM; 2012.

7. Loos A, Ernst A: Detection and identification of chimpanzee faces in the wild. In *IEEE International Symposium on Multimedia*. CA: Irvine; 10–12 December 2012.

8. Loos A: Identification of primates using global and local features. In *IEEE International Conference on Acoustics, Speech, and Signal Processing (ICASSP)*. Vancouver; 26–31 May 2013.

9. Ernst A, Küblbeck C: *Fast face detection and species classification of African great apes, 8th IEEE International Conference on Advanced Video and Signal Based Surveillance*. New York: IEEE; 2011.

10. Bay H, Ess A, Tuytelaars T, Gool LV: Speeded-up robust features (SURF). *Comput. Vis. Image Underst* 2008, 110(3):346-359.

11. Rowley HA, Baluja S, Kanade T: Neural network-based face detection. *IEEE Trans. Pattern Anal. Mach. Intell*1998, 20: 23-38.

12. Viola P, Jones M: *Rapid object detection using a boosted cascade of simple features, IEEE Conference on Computer Vision and Pattern Recognition, vol. 1*. New York: IEEE; 2001.

13. Freund Y, Schapire RE: A short introduction to boosting. *J. Japanese Soc. Artif. Intell* 1999, 14(5):771-780.

14. Viola P, Jones M: Robust real-time object detection. *Int. J. Comput. Vis* 2002, 57(2):137-154.

15. Lienhart R, Maydt J: *An extended set of haar-like features for rapid object detection, IEEE International Conference on Image Processing (ICIP), vol. 1*. New York: IEEE; 2002.

16. Wu B, Haizhou A, Chang H, Shihong L: *Fast rotation invariant multi-view face detection based on real Adaboost, 6th IEEE International Conference on Automatic Face and Gesture Recognition*. New York: IEEE; 2004.

17. Wawerla J, Marshall S, Mori G, Rothley K, Sabzmeydani P: BearCam: automated wildlife monitoring at the Arctic Circle. *Mach. Vis. Appl* 2009, 20(5):303-317.

18. Burghardt T, Calic J: Real-time face detection and tracking of animals. In *8th Seminar on Neural Network Applications in Electrical Engineering*. Belgrade: Serbia & Montenegro; 25–27 September 2006:27-32.

19. Spampinato C, Giordano D, Di Salvo R: Automatic fish classification for underwater species behavior understanding. In *ACM International Workshop on Analysis and Retrieval of Tracked Events and Motion in Imagery Streams (ARTEMIS)*. Italy: Firenze; 29 October 2010:40-50.

20. Spampinato C, Palazzo S, Boom B, van Ossenbruggen J, Kavasidis I, Di Salvo R, Lin FP, Giordano D, Hardman L, Fisher RB: Understanding fish behavior during typhoon events in real-life underwater environments.*Multimedia Tools Appl* 2012. http://scholar.google.com/citations?view_op=view_citation%26hl−de%26user=yIr6TqAAAAJ%26citation_for_view=yJr6TqAAAAAJ:qjMakFHDy7sC

21. Turk M, Pentland A: Eigenfaces for recognition. *J Cogn. Neurosci* 1991, 3: 71-86.

22. Belhumeur PN, Hespanha JP, Kriegman DJ: Eigenfaces vs. Fisherfaces: recognition using class specific linear projection. *IEEE Trans. Pattern Anal. Pattern Anal. Mach. Intell* 1997, 19(7):711-720.

23. He X, Yan S, Hu Y, Niyogi P, Zhang HJ: Face recognition using Laplacianfaces. *IEEE Trans. Pattern Anal. Mach. Intell* 2005, 27(3):328-40.

24. Wright J, Yang AY, Ganesh A, Sastry SS, Ma Y: Robust face recognition via sparse representation. *IEEE Trans. Parallel Patt. Anal. Mach. Intell* 2009, 31(2):210-27.

25. Yang M, Zhang L: Gabor feature based sparse representation for face recognition with Gabor occlusion dictionary, vol. 6316. In *European Conference on Computer Vision (ECCV), Lecture Notes in Computer Science*. Heidelberg: Springer; 2010:448-461.

26. Ardovini A, Cinque L, Sangineto E: Identifying elephant photos by multi-curve matching. *Pattern Recognit* 2008, 41(6):1867-1877.

27. Araabi BN, Kehtarnavaz N, McKinney T, Hillman G, Würsig B: A string matching computer-assisted system for dolphin photoidentification. *Ann. Biomed. Eng* 2000, 28(10):1269-1279.

28. Burghardt T, Campbell N: Individual animal identification using visual biometrics on deformable coat patterns. In*5th International Conference on Computer Vision Systems (ICVS)*. Bielefeld; 21–24 March 2007. http://biecoll.ub.uni-bielefeld.de/volltexte/2007/20/

29. Burghardt T: Visual animal biometrics - automatic detection and individual identification by coat pattern,. PhD Thesis, University of Bristol, 2008.

30. Lahiri M, Warungu R, Rubenstein DI, Berger-Wolf TY, Tantipathananandh C: *Biometric animal databases from field photographs: identification of individual zebra in the wild, ACM International Conference on Multimedia Retrieval (ICMR)*. New York: ACM; 2011.

31. Schapire RE, Singer Y: Improved boosting algorithms using confidence-rated predictions. *Mach. Learn* 1999, 37(3):297-336.

32. Fröba B: *Verfahren zur Echtzeit-Gesichtsdetektion in Grauwertbildern*. Aachen: Shaker; 2003.

33. Zabih R, Woodfill J: Non-parametric local transforms for computing visual correspondence. In *Third European Conference on Computer Vision Proceedings, Volume II*. Stockholm, Sweden; 2–6 May 1994:151-158.

34. Ojala T, Pietikäinen M, Harwood D: A comparative study of texture measures with classification based on featured distributions. *Pattern Recognit* 1996, 29: 51-59.

35. Gao Y, Wang Y, Feng X, Zhou X: *Face recognition using most discriminative local and global features. International Conference on Pattern Recognition (ICPR)*. New York: IEEE; 2006.

36. Wiskott L, Fellous JM, Krüger N, von der Malsburg C: Face recognition by elastic bunch graph matching. *IEEE Trans. Pattern Anal. Mach. Intell.* 1997, 19(7):775-779.

37. Liu C, Wechsler H: Gabor feature based classification using the enhanced fisher linear discriminant model for face recognition. *IEEE Trans. Image Proc* 2002, 11(4):467-476.

38. Xie S, Shan S, Chen X: Fusing local patterns of Gabor magnitude and phase for face recognition. *IEEE Trans. Image Proc* 2010., 19(5):

39. Lowe D: Distinctive image features from scale-invariant keypoints. *Int. J. Comput. Vis* 2004, 60(2):91-110.

40. Chang CC, Lin CJ: LIBSVM: a library for support vector machines. *ACM Trans. Intell. Syst. Tech* 2011., 2: http://dl.acm.org/citation.cfm?id=1961199.

41. Gökberk B, Salah AA, Akarun L: *Rank-based decision fusion for 3D shape-based face recognition. International Conference on Audio- and Video-Based Biometric Person Identification (AVBPA)*. Berlin Heidelberg: Springer-Verlag; 2005.

42. Wu TF, Lin CJ, Weng R: Probability estimates for multi-class classification by pairwise coupling. *J Mach. Learn. Res* 2004, 5: 975-1005.

43. Ernst A, Ruf T, Küblbeck C: A modular framework to detect analyze faces for audience measurement systems. In *Proceedings of the 2nd Workshop on Pervasive Advertising 2009 In conjunction with Informatik 2009*. Germany: Lübeck; 2 October 2009:3941-3953.

44. Phillips PJ, Rizvi S, Rauss P: The FERET evaluation methodology for face recognition algorithms. *IEEE Trans. Pattern Anal. Mach. Intell* 2000, 22: 1090-1104.

45. Phillips PJ, Grother P, Micheals R, Blackburn D, Tabassi E, Bone J: Face recognition vendor test 2002: evaluation report. (Nistir 6965, National Institute of Standards and Technology, 2003)

46. Phillips PJ, Grother P, Micheals R: Chapter 21: evaluation methods in face recognition. In *Handbook of Face Recognition*. London: Springer-Verlag; 2011:551-57

47. Ahumada JA, Silva CEF, Gajapersad K, Hallam C, Hurtado J, Martin E, McWilliam A, Mugerwa B, O'Brien T, Rovero F, Sheil D, Spironello WR, Winarni N, Andelman SJ: Community structure and diversity of tropical forest mammals: data from a global camera trap network. *Philos. Trans. R. Soc. London B* 2011, 366(1578):2703-2711.

48. Head J, Boesch C, Robbins M, Rabal L, Makaga L, Kuehl H: Effective socio-demographic population assessment of elusive species for ecology and conservation management. *Ecol. Evol* 2013. http://onlinelibrary.wiley.com/doi/10.1002/ece3.670/abstract 10.1002/ece3.670

49. Hoppe-Dominik B, Kühl H, Radl G, Fischer F: Long-term monitoring of large rainforest mammals in the biosphere reserve of Taï National Park, Côte d'Ivoire. *Afr. J. Ecol* 2011, 49(4):450-458.

Chapter 12

Novel Solution Based on Face Recognition to Address Identity Theft and Cheating in Online Examination Systems

Ayham Fayyoumi[1], Anis Zarrad[2]

[1]Department of Information Systems, Al Imam Mohammad IBN Saud Islamic University (IMSIU), Riyadh, Saudi Arabia

[2]Department of Computer Science and Information Systems, Prince Sultan University (PSU), Riyadh, Saudi Arabia

ABSTRACT

The main objective of this research is to provide a solution for online exam systems by using face recognition to authenticate learners for attending an online exam. More importantly, the system continuously (with short time intervals), checks for learner identity during the whole exam period to ensure that the learner who started the exam is the same one who continued until the end and prevent the possibility of cheating by looking at adjacent PC or reading from an external paper. The system will issue an early warning to the learners if suspicious behavior has been noticed by the system. The proposed system has been presented to eight e-learning instructors and experts in addition to 32 students to gather feedback and to study the impact and the benefit of such system in e-learning environment.

KEYWORDS

Online Exam, Face Recognition, Authentication, Exam Cheating

1. INTRODUCTION

Nowadays, e-learning systems have become vital components in the education and training domains. Several countries are attempting to overcome the Knowledge Divide. Through education and training, countries are able to develop the skills of their citizens, consequently bridging the Knowledge Divide within the country and with more developed ones. Success in the Knowledge Economy relies heavily on a qualified and skilled population, thus effective education and training systems are required.

Simultaneously, Information and Communication Technologies (ICTs) continue to grow at a rapid pace and have changed the way people live, work, and learn. The integration of ICT tools in education and training has created new ways of delivering, accessing, and processing useful knowledge, as well as has provided support to knowledge sharing between different actors and to lifelong learning [1] . In addition, technological development and the growth of the Internet have resulted in the emergence of e-learning as an important learning approach. E-learning provides innovative methods for educating people. Moreover, the e-learning market is expanding because of its many advantages over traditional education. E-learning is also highly flexible, scalable, employs a rapid learning method, less expensive, and proven to be effective compared with traditional education. In particular, the following are the three main drivers for the increasing global importance of e-learning:

- Movement toward a knowledge-based economy;
- Paradigm shift in education delivery;
- Technological developments and Internet growth.

The development of e-learning and online assessment systems is increasing rapidly, both globally and locally, with many universities and corporations investing significant capital in e-learning programs and initiatives. This growth is also reflected in the report by Ambient Insight, which was published in 2010, indicating that the e-learning market has reached US$ 27.1 billion in 2009 and will surpass $49.6 billion by 2014 [2] . The growth of the e-learning industry requires new services to ensure reliability and effectiveness of its systems, especially during the examinations process, by addressing the issue of cheating in online examinations and identity theft.

E-learning is prospering on global and local levels. In Saudi Arabia, the government is focusing to the education sector in general and to e-learning in particular in responding to the increasing number of male and female students enrolled in educational institutions. Many universities in Saudi Arabia have already implemented e-learning systems and are offering distance learning courses and degrees. Thus, ensuring the reliability of e-learning systems, especially during examinations, is highly critical. Online examination cheating and identity theft should be considered, while the privacy of learners' data and more importantly, their images is guaranteed.

2. E-LEARNING SYSTEMS

The Web is easy to use, easy to update, and is available worldwide. The Web is the driver of the knowledge economy and is therefore a natural vehicle for learning [3] . Hall and Snider defined e-learning as synonymous to all computer-related applications, tools, and processes of learning and teaching [4] .

E-learning offers flexible learning anytime and anywhere. The increasing speed of the Internet, the growth of the World Wide Web, and the emergence of high-speed computers contribute to the availability of e-learning 24/7 and worldwide. Moreover, e-learners can access materials at any time and place convenient for them. E-learning has other advantages for learners, teachers, and instructional developers. Learners, for example, will benefit from their interaction with other students and with their instructors, and they can study at their preferred pace. Learners can also select the material they want or be directed to the content that meets their level of knowledge, interest, and needs. Furthermore, learners become responsible for their learning. Meanwhile, teachers can develop materials using online resources, and then publish them in many different ways such as text, images, video, audio, simulations, and games. Teachers will subsequently gain satisfaction through quality student participation. Finally, developers can develop detailed and standardized courses. E-learning allows developers to design the course and use it for multiple times by using learning objects.

E-learning provides much enrichment to the many models, markets, interest groups, and different degrees of satisfaction in the educational process. E-learning technologies offer a potential for high-quality formative assessment. Online assessment provides dynamic visuals, sound, user interactivity, adaptation to individual learners, and almost real-time score reporting, thus expanding examination options beyond the limitations of traditional tests [5] .

Learner assessments are undoubtedly essential in the educational process. Examination scores inform the instructor on whether a student's progress is satisfactory or not. Online assessment systems provide instructors with many advantages, such as the creation of online examinations within a short time, administration of the examination through the computer, easy monitoring of answers during the examination, fast access to examination results without spending time on evaluation and correction, and easy calculation of the trends of the examination results. In addition, online examinations benefit learners by allowing them to not be physically present at a given location to take an examination, and the results can be made available to them immediately.

In online examinations, learners should register their names and passwords. The examination items will then be generated by the test bank according to the parameters set by the instructor. The items in online assessments are usually true/false and multiple choice questions, as well as involves reordering/rearrangement (matching, categorization, ranking, and others), completion, concept maps, and essays. Examination questions will appear on the screen, and then each learner will start answering them through his/her computer. At the end of the time limit, the examination will stop and the score will appear. The learner will then obtain the test results immediately.

3. ONLINE ASSESSMENT RELIABILITY

Security issues of e-learning systems have been discussed by many researchers [6] . Online examination is a challenge for e-learning security [7] . Currently, online assessments are mostly conducted at specific examination centers and require supervision mainly because, if administered in unsupervised locations, learners may acquire assistance from others to improve their examination results or have another individual take the examination for them. In such cases, instructors will become uncertain on who answered the examination questions, which conflicts the flexibility advantage of online education. Therefore, despite the expansion of the e-learning market locally and globally, a problem remains, especially with off-site examinations. Failure to verify the learners attending an examination is a major challenge in online learning environments.

Very little attention has been given in solving the problem of learners' unethical conduct. Moreover, many researchers criticized current e-learning systems for not focusing on the authentication of learners, particularly those who engage in online quizzes and examinations [8] . McGinity noted that biometrics has replaced the conventional password systems [9] . Another study highlighted the importance of detection mechanisms beyond the initial access to the e-learning system [10] . Therefore, a system must be developed to ensure that the person taking an examination is the student enrolled in the course. Yang and Verbauwhede suggested that biometrics systems provide better security than password systems [11] . Moreover, Hugl highlighted many technologies related to security that have not been used in e-learning [12] . These technologies include biometrics technologies that are increasingly becoming essential in a number of applications. Biometric authentication is the automatic recognition and identification of learners by using their physiological characteristics such as voice, hand geometry, fingerprints, and facial images. Generally, biometric authentication requires comparison of the stored data against the captured data.

No perfect biometric system that fits all needs has so far been created. All known systems have their advantages and disadvantages. A few studies had focused on improving e-learning security using biometric systems, but a limited number of them addressed the issue of continuous user authentication. In a recent study, Flior and Kowalski discussed a method for providing continuous biometric user authentication in online examinations via keystroke dynamics [13] . However, keystroke biometrics has its disadvantages, such as the major differences that can occur over time as a result of changes in typing pattern, tiredness of the hands after a period of typing, and improvement of skills.

In line with the previously mentioned concept, researchers are presently looking for the best biometric authentication method that will help validate the identification of the learner attending the examination and that will ensure that he/she is the same person as the one registered in the course without compromising his/her privacy. Face recognition systems are human-friendly because they require no contact and no additional hardware (given that most PCs and laptops come with a camera). More importantly, face recognition systems can be used for the continuous authentication of the learner during the entire examination period.

4. PROTOTYPE DEVELOPMENT: THE PROPOSED SOLUTION

Face recognition technologies operate by scanning the person's face and matching it with the stored image. The face recognition biometric system is a system that records distinguished facial features and stores the template in a server. In scanning the face, the camera identifies facial features and transmits the signal to the server where the scanned features are processed for matching (See figure 1). Facial recognition identifies key features from the facial image. The system detects the face and captures image(s) of facial features that do not change over time, while avoiding those that change, such as facial expressions or hair. The first step in face recognition is the detection of the face in the image. Yang and his colleagues evaluated the main methods used for face detection, namely, knowledge-based method, appearance-based approach, feature invariant method, and template-matching method. The second step is the conduct of several approaches to model and recognize the facial image, such as direct correlation, elastic graph matching, neural networks, principal component analysis (PCA), and multiresolution analysis [14] .

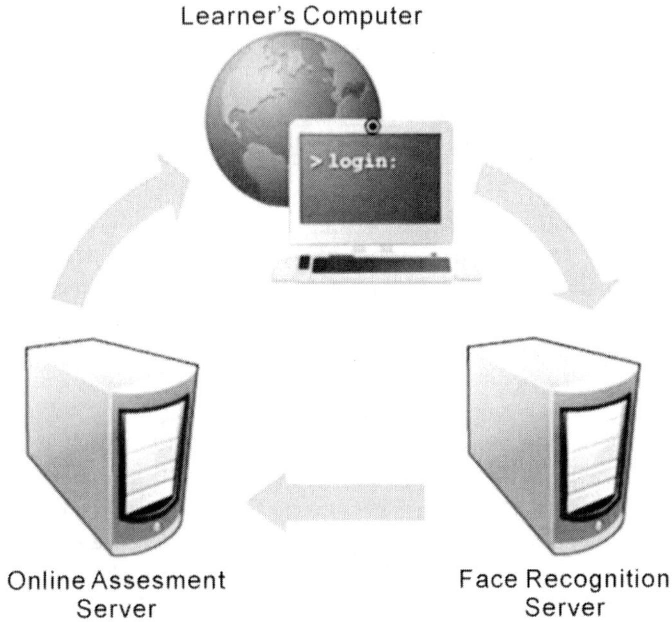

Figure 1. Proposed system solution.

The proposed examination system includes the development of a test bank on a specific field. The test bank contains a variety of question items (e.g., matching, ranking, essay, categorization, multiple-choice items, and true/false questions). These questions are classified into three difficulty levels: easy, moderate, and difficult. In designing the test, the instructor can specify the number of test questions, types of questions, and the difficulty associated to each question. The system will then automatically generate a random set of questions based on the criteria specified by the instructor. Therefore, each learner will receive a different set of questions but with the same difficulty level. The instructor is also allowed to design a new question and specify its difficulty. The newly formulated questions will be stored in the database and then added into the test bank. At the end of the examination time or after completing the test, the score and time spent by each learner during the examination will appear. The database will also save other data related to the registered users or learners, including profiles and images.

The face recognition system is integrated in the online assessment tool to identify and verify the learners allowed to access the exam and to continuously validate the learner's identity until the end of the examination. Specifically, during registration in the course (when used as part of an e-learning system) or for an examination, images of learners, in addition to other required data, are captured and stored in the database. Captured images are encrypted to protect learners' privacy. During examination time, the learner's identity is verified for attendance in the examination and is monitored by comparing the captured images with the one stored in the database. Figure 2 shows an overview of the proposed system architecture.

In addition, to address the issue of cheating in online examination systems, continuous checking is implemented, as shown in Figure 3. In the said figure, the learner is captured looking at the screen (left), reading (middle), and looking at an adjacent learner's PC (right). In the two-second video taken during the examination period, the images in the video are compared with each other to verify if the learner was looking somewhere else other than his/her screen. If, for all images within the two seconds, the learner was not looking at the screen and therefore was not focused in solving the examination questions, he/she will be warned by a change in background color, as shown in Figure 4. Failed authentication will also be made visible to the learners by the change in background color.

If the authentication failure continues for more than a few seconds, the system will stop and perform collaborative verification. In this stage, the system will ask the user to put his/her face in an appropriate position to capture a new image. If the error is repeatedly encountered, the examination will not be administered for suspected cheating.

In practice, for learners not to be disturbed and to minimize matching failures, a two second video of the learner is recorded and the best image in terms of facial expressions, lighting, and resolution will be selected by an application previously installed on the client's side of the system. This image is sent to the server and will be compared with the stored image.

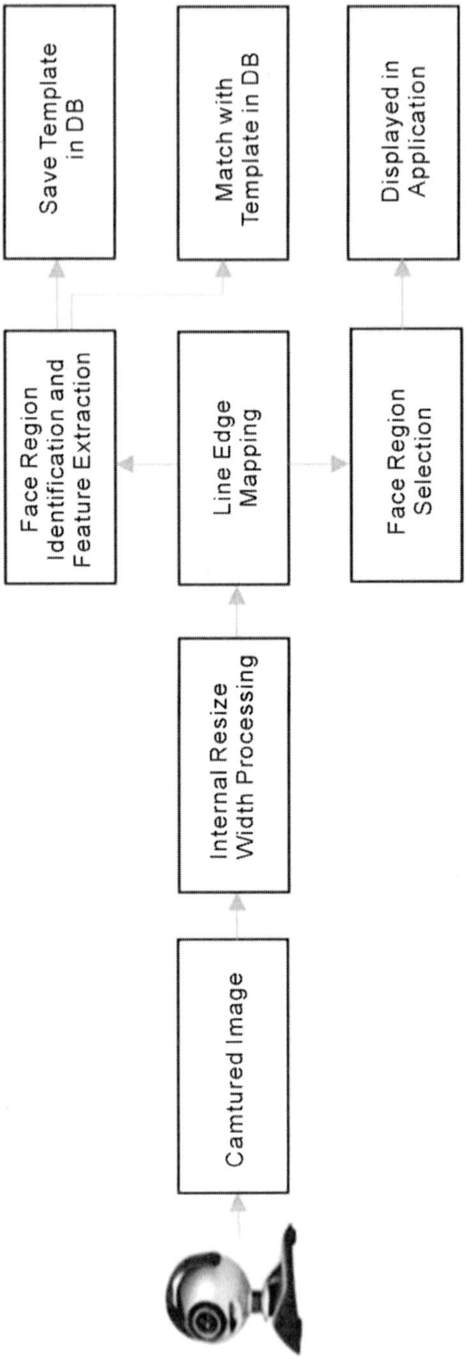

Figure 2. Face recognition system architecture.

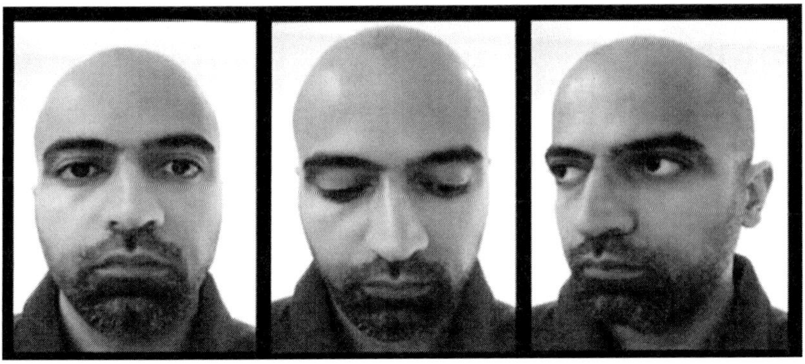

Figure 3. Images of learner at different positions.

5. SURVEY RESULTS

Two surveys were conducted. The first survey targeted online instructors, and the second targeted potential users/students. The first investigation was conducted to identify the rate of image capture and the number of times the suspicious behavior is accepted without affecting learners' concentration during the examination. Moreover, e-learning instructors and experts (n = 8) were provided with the description of the proposed solution and were asked to rate the following statements using a five-point Likert scale (5—Strongly Agree to 1—Strongly Disagree): 1) the system will provide reliable results that reflect learners' achievements, 2) the system will effectively validate the learners' identity, 3) the probability of cheating during examinations will be minimized, 4) I will recommend this system to other instructors, and 5) the students will study harder. In addition, the instructors and experts were asked to specify the following information: 6) the capture rate of images for presence checks and 7) the acceptable number of times of suspicious behavior (warnings) during the examination. Figure 5 shows the results of the survey involving e-learning instructors and experts.

Table 1 presents the suggested rate of image capture and the suggested acceptable number of warnings.

Meanwhile, learners were engaged in an online quiz in a Blackboard environment and were provided with an explanation of the proposed system. The learners (n = 32) were asked to rate the following statements: 1) the probability of cheating during exams will be minimized, 2) the system's operation is easy to understand, 3) the system will benefit the student (e.g., encourage more focus in answering the examination), and 4) the system will encourage students to study harder.Figure 6 shows the results of this survey involving learners.

6. CONCLUSIONS

Results show that almost all e-learning instructors agree with the given statements. However, some instructors expressed doubts regarding the effective authentication of learners and the reduced probability of cheating during examinations. Meanwhile, the students' survey showed a high percentage of neutral cases, mainly regarding the benefits that the system will provide to students.

Two questions were asked to both e-learning instructors and students. These questions were whether the probability of cheating will decrease and whether the system will encourage students to study harder. Both parties believed that cheating will be reduced with the implementation of the facial recognition system and will motivate students to study harder. The facial recognition system should consider the image capture rate of 30 seconds for presence checks. This balanced value is appropriate to lessen the load on the system and to guarantee the level of reliability. Moreover, the learner with suspicious behavior should be warned twice.

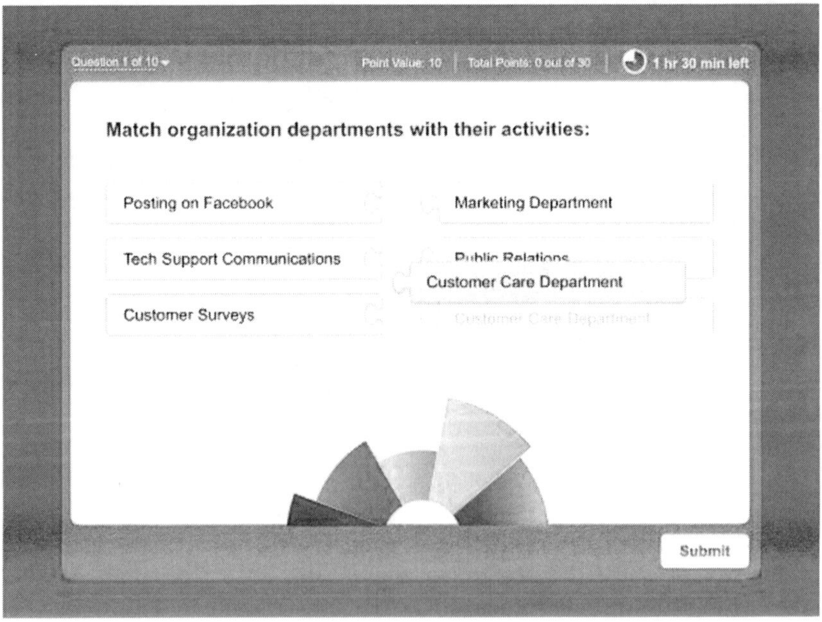

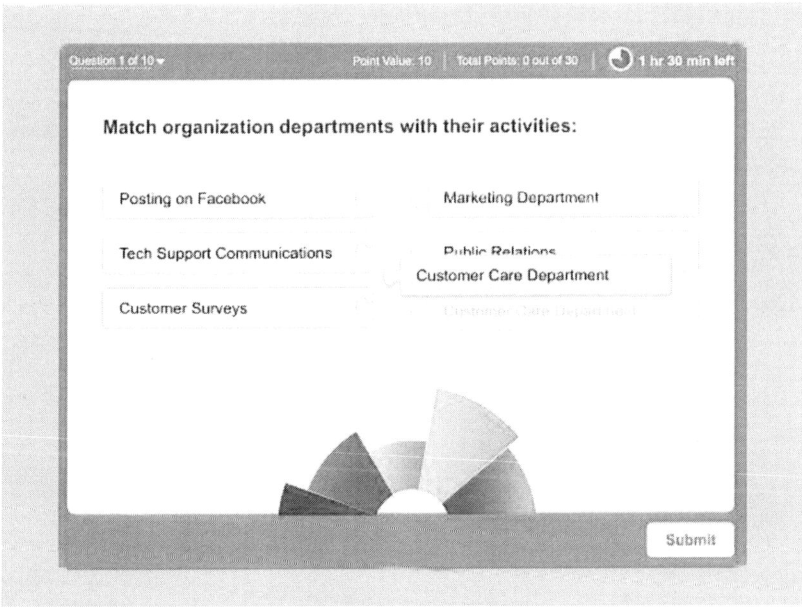

Figure 4. Red color as warning for learner with suspicious behavior.

Table 1. Suggested values from e-learning instructors/experts.

	Every 10 sec	Every 30 sec	Ever 1 min	Every 5 min	Every 10 min
The rate of images captured for presence checks	1	3	3	1	0
	1 time	2 times	3 times	4 times	5 or more
The number of times suspicious behavior is accepted during the exam	1	4	2	1	0

Other applications of the facial recognition system include monitoring the attendance in virtual classrooms. In measuring the attendance rate, if the result of the student's verification is positive, the system will register an attendance mark for him/her. Otherwise, he/she will be marked absent.

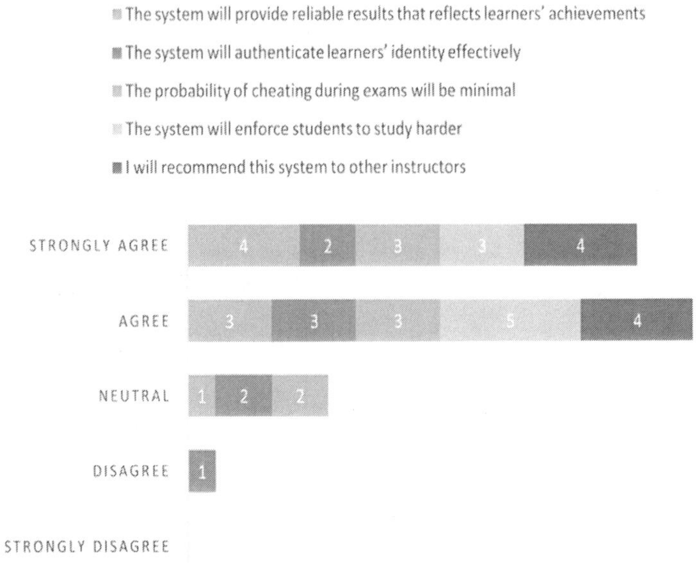

Figure 5. Results of e-learning instructors/experts survey.

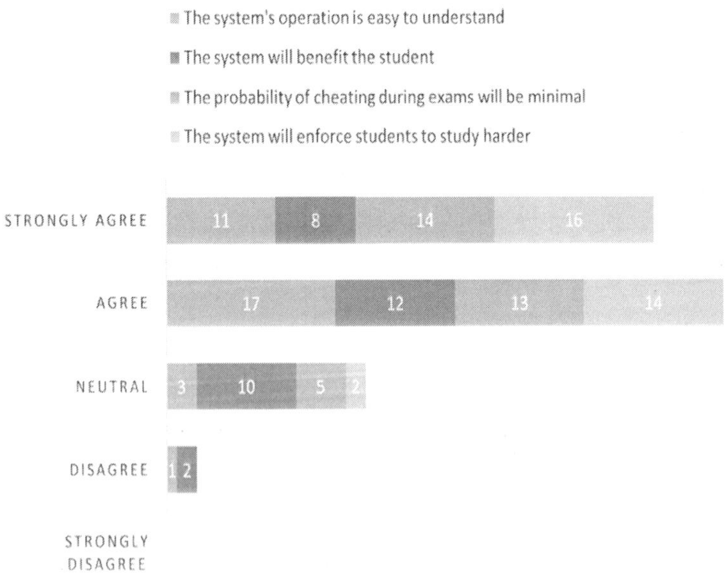

Figure 6. Results of e-learners survey.

REFERENCES

1. Assaf, W., Elia, G., Fayyoumi, A. and Taurino, C. (2007) Prospect of e-Learning: The Case of Jordan. e-Society 2007 —IADIS Multi Conference on Computer Science and Information Systems, Lisbon, 3-8 July 2007.

2. Ambient Insight (2010) The Worldwide Market for Self-Paced eLearning Products and Services: 2009-2014 Forecast and Analysis Report.

3. Rosenberg, M.J. (2001) E-Learning: Strategies for Delivering Knowledge in the Digital Age. McGraw-Hill, New York.

4. Hall, B. and Snider, A. (2000) Glossary: The Hottest Buzz Words in the Industry.

5. Scalise, K. and Gifford, B. (2006). Computer-Based Assessment in E-Learning: A Framework for Constructing "Intermediate Constraint" Questions and Tasks for Technology Platforms. Journal of Technology, Learning, and Assessment, 4.

6. Levy, Y. and Ramim, M. (2007) A Theoretical Approach for Biometrics Authentication of e-Exams. Nova Southeastern University, 93-101.

7. Huszti, A. and Petho, A. (2008) A Secure Electronic Exam System. Informatikafelsőoktatásban, 1-7.

8. Huang, W., Yen, D. C., Lin, Z.X. and Huang, J.H. (2004) How to Compete in a Global Education Market Effectively: A Conceptual Framework for Designing a Next Generation eEducation System. Journal of Global Information Management, 12, 84-107.

9. McGinity, M. (2005) Staying Connected: Let Your Fingers Do the Talking. Communications of the ACM, 48, 21-23.

10. Pillsbury, C. (2004) Reflections on Academic Misconduct: An Investigating Officer's Experiences and Ethics Supplements. Journal of American Academy of Business, 5, 446-454.

11. Yang, S. and Verbauwhede, I. (2003). A Secure Fingerprint Matching Technique. Proceedings of the 2003 ACM SIGMM workshop on Biometrics Methods and Applications, California, 89-94.

12. Hugl, U. (2005) Tech-Developments and Possible Influences on Learning Processes and Functioning in the Future. Journal of American Academy of Business, 6, 250-256.

13. Flior, E. and Kowalski, K. (2010) Continuous Biometric User Authentication in Online Examinations. Seventh International Conference on Information Technology IEEE Computer Society, Las Vegas, 12-14 April 2010, 488-492.

14. Yang, M., Kriegman, D. and Ahuja, N. (2002) Detecting Faces in Images: A Survey. IEEE Transactions on Pattern Analysis and Machine Intelligence, 24, 34-58.

Index

L

M

O

P

S

T

U

V

W